THE
FUTURE
IS
NOW

Also by Bob McDonald

An Earthling's Guide to Outer Space:
Everything You Ever Wanted to Know about
Black Holes, Dwarf Planets, Aliens, and More

Canadian Spacewalkers:
Hadfield, MacLean and Williams
Remember the Ultimate High Adventure

Measuring the Earth with a Stick:
Science as I've Seen It

Wonderstruck II

Wonderstruck

SOLVING THE CLIMATE CRISIS
WITH TODAY'S TECHNOLOGIES

THE FUTURE IS NOW

BOB McDONALD

Host of CBC's *Quirks & Quarks*

VIKING

VIKING

an imprint of Penguin Canada, a division of Penguin Random House Canada Limited
Canada • USA • UK • Ireland • Australia • New Zealand • India • South Africa • China

First published 2022

LIBRARY AND ARCHIVES CANADA CATALOGUING IN PUBLICATION

Title: The future is now : solving the climate crisis with today's technologies/ Bob McDonald.
Names: McDonald, Bob, 1951- author.
Identifiers: Canadiana (print) 20210395702 | Canadiana (ebook) 20210395710 |
ISBN 9780735241947 (hardcover) | ISBN 9780735241954 (EPUB)
Subjects: LCSH: Renewable energy sources—Popular works. |
LCSH: Clean energy—Popular works.
Classification: LCC TJ808 .M33 2022 | DDC 333.79/4—dc23

Book design by Dylan Browne
Cover design by Talia Abramson
Cover images: (circuit tree) © ex_artist, (leaves) © Mallinka1, both Shutterstock.com

Printed in Canada

10 9 8 7 6 5 4 3 2

Penguin
Random House
VIKING CANADA

To all who see a clear path to a green energy future

"Rise above oneself and grasp the world."
—*Archimedes*

Contents

Let's start with the good news. The technology to produce energy without carbon emissions already exists. There is as much solar energy beaming down from the sky onto the Earth every hour that humanity consumes in a year.[1] Energy blows on the wind, boils out of the ground, and literally grows on trees. There is thousands of times more non-fossil fuel energy available than even our hungry consumer mouths are swallowing. And we have the means to gather it up and put it to useful work.

The challenge is not just constructing the wind turbines, solar farms, geothermal plants, tidal stations, biofuels and other alternative energy devices, the big elephant in the room is giving up the most convenient, versatile source of energy that has driven civilization since the Industrial Revolution and turned humans into a superspecies. Black gold.

Fossil fuels are incredible. They pack an amazing amount of energy into a very small space. Gasoline has one hundred times the energy density of lithium-ion batteries, and the batteries are about fifty times heavier. Fossil fuels are easy to carry around, can remain in storage without losing energy until we need them, and, besides providing the power to keep trains, planes, and automobiles on the move, can be broken down into many useful products, from plastics to synthetic fabrics, fertilizer, even coffee whitener. No wonder they're so popular.

Too bad that the way we burn them is changing the climate.

Alternatives to fossil fuel energy are everywhere, but clean sources such as solar and wind are spread over large areas, so it takes large technology covering a lot of land to gather it up and concentrate it into useful forms. Fossil fuels come out of the ground already densely packed with energy, which is why they have been the fuel of choice for centuries. And why we burn a lot of it.

Our thirst for oil is insatiable. Worldwide, we consume around 100 million barrels of it every *day*.[2] That's 4 billion gallons, 15 billion litres, or enough to keep Niagara Falls running for two hours. That's just one day of quenching our thirst for oil. And that consumption continues to rise as both our population and the popularity of new energy-consuming devices grow. Of all the fossil fuels consumed since the beginning of the Industrial Revolution about 150 years ago, more than half has been used in the last 50 years.[3] All we had to do was dig this black gold out of the ground and light a match to reap its many benefits.

Look around you. Everything that powers up, gets warm, or moves is using energy. From the alarm clock that wakes you in the morning to the lights in your home, the heat to cook meals, and the vehicles to transport you. And we assume all that energy will always be there. Flip a switch and lights come on. Plug in a device and it automatically charges up. Turn up the thermostat and your home becomes warmer. Press the accelerator pedal and your vehicle moves. And for the most part, that energy is invisible. We don't see the electricity that came from a generating station and into the wires of your home, the gas in the pipelines that feed a furnace, or the combustion inside an engine that turns the wheels.

We also don't see where that energy comes from, whether from burning coal, oil, or natural gas in generating stations, water falling through the penstocks of a dam, or the nuclei of atoms smashing into one another in a nuclear reactor. All these forms of energy are simply there when we need them, and they have become so entrenched in our lifestyles that we can't live without them. The only time we get a sense of how much we consume is when the monthly bills come in and it hits us in the pocketbook. But trying to quantify just how much energy all of humanity consumes is as difficult as trying to count the stars in the universe. The numbers are incredible.

According to the International Energy Agency (IEA), our global energy demand from all sources is about 120 million tons of oil equivalent, known as Mtoe, every year.[4] That's a lot (a gross under-statement). We use it in such quantities because oil contains a huge amount of energy. One 45-gallon drum, seen being used by the military or shined up and played as an instrument in a Jamaican steel band, is a standard barrel, or 159 litres. To measure the energy in a barrel, another unit is used, the metric joule, which is how much work that energy can do. One joule is the energy used to lift 1 kilogram up 1 metre. So pick up a small bag of oranges from the floor and your muscles are burning one joule of energy.

A single barrel of oil holds 6 billion joules. That's right, *6 billion*. So how much work could that oil do?

Think about the great pyramids of Egypt, the largest manmade monuments on the planet. Based on the mass of all the stones that make up the pyramids and the height to which those stones were lifted during construction, in pure energy terms, it has been calculated that it took about 2.4 trillion joules to build those ancient structures.[5] Don't tell that to the slaves who dragged those stones around—they would say they put out that many gallons of sweat. Translated into oil, the pyramids could be built with about 400 barrels. That's not a lot of oil. A typical oil well can pump that amount in one day. That is less than one second of world oil production. So when we look at our total global energy consumption from all sources—fossil fuels, hydro, geo-thermal, wind, solar, nuclear—as 120 million tons of oil equivalent, it works out to more than two million pyramids every year!

It took the Ancient Egyptians twenty years to build them once.

Our energy consumption goes beyond keeping ourselves warm, well lit, and moving from place to place. Vast amounts are consumed mining minerals out of the ground and heating blast furnaces to smelt ore into metals. We literally move mountains and change the course of rivers to cover the land in concrete and asphalt, then build towers of glass and steel above them. We manufacture products and ship them around the globe. Northern countries are supplied with fresh fruit and vegetables during winter months thanks to a huge trans-portation network that brings food up from the tropics. Everywhere we look, energy is being used in one form or another. Just a few city blocks of downtown Manhattan consume a pyramid's worth of energy every day. If the world population continues to grow to nine billion, and fossil fuel use maintains business as usual, we will not reach the target of limiting climate increase to 1.5 C.[6]

These are staggering numbers, and the only reason we have been able to become such energy hogs is because fossil fuels are abundant, and we have become very good at digging them out of the ground and shipping them around the globe. That, and the fact that they carry so much energy in such a small portable space, giving them that pyramid-building potential, means they still dominate the energy we consume.

Is it any wonder that fossil fuels powered the Industrial Revolution? Think about the convenience of a lump of coal. It is a rock that burns. Just dig it up, light a match, and boom, instant heat. Use it to boil water and you have steam power to run factories and drive locomotives. So much punch in such a small package. Oil has even more punch, and it is fairly easy to handle. Fuels can be stowed in storage tanks, coal bunkers, or underground reservoirs. If we need instant energy, or want to go somewhere in a hurry, just shovel coal into the firebox or pour liquid fuel into a gas tank, and we can cross the country. And when we need more, there is always a refuelling station close at hand. Yes, fossil fuels and the network that supplies them are amazing.

Unfortunately, throughout most of the history of burning, no one thought about the products of combustion that were spewing out of smokestacks and tailpipes. They simply blew away on the wind, out of sight, out of mind. Now we know that those gases behave like the glass of a greenhouse, and are warming the climate.

But the fossil fuels themselves are not the problem. It is the inefficient ways we have been burning them that is changing the chemistry of the atmosphere. The internal combustion engines that have been powering vehicles for more than a century are around 20 percent efficient.[7] (Efficiency varies depending on the size of a vehicle, type of engine, and body shape. A big, heavy, four-wheel-drive truck with a large engine and boxy shape is less efficient than a small lightweight vehicle with a smooth aerodynamic shape.) Efficiency is converting energy from one form to another, and if the amount of chemical energy put into a system—say, the energy in gasoline—is the same as the amount of mechanical energy coming out—such as the turning of the wheels in a car—then the system is 100 percent efficient. But no energy conversion is that good. There is always energy lost along the way. Even if you drive a small car that you think gets great mileage, only 20 percent of the energy in the gasoline ends up driving the wheels. The rest of the energy is lost as heat thrown out the tailpipe or carried away by the radiator. Then there are losses due to friction from all the moving parts within the engine, rolling resistance on the highway, and pushing the wind out of the way, although all vehicles have to deal with the last two. In other words, a combustion engine is 80 percent inefficient. But thanks to that supreme energy density of gasoline, we can afford to throw most of it away and still have enough to drive a full-sized vehicle at high speed down the highway for hundreds of kilometres. We accept that as good mileage. But think of it this way: If you spend $50 filling up your car, only $10 is used to move it. You are essentially throwing away $40.

Imagine going to a filling station, pumping gas into the fuel tank but stopping at $10. Then remove the nozzle from the car and spray $40 of gas into the air. You would probably be arrested for polluting, and it would look like you are wasting gas. But when you burn gas in the engine, most of the products of that burning are indeed thrown into the air. Waste heat, of course, but also combustion products such as carbon dioxide and carbon monoxide that remain in the atmosphere. An average vehicle will produce more than its own weight in carbon every year. Such is the hard reality of heat engines and technologies that burn fossil fuels directly.

Improving efficiency would go a long way towards reducing our impact on the climate and, while we're at it, finding other ways to get energy out of fossil fuels without all those nasty by-products. New research is looking into cleaner, more efficient ways to get energy out of oil, such as extracting hydrogen, which is then run through a fuel cell to make electricity and drive an electric car. That is just one way the oil industry can remain viable.

Fossil fuels have transformed humans into a superspecies, enabling us to spread over the entire planet, including places humans don't normally belong, such as flying high through the air or diving to the bottom of the ocean. We have extended our lifespans through better food and health care, built a globe-circling communications network, sent robots to every planet in the solar system, with humans soon to follow. The Earth has seen many giant species come and go over time, but there has never been anything like *Homo sapiens*. We have eliminated our natural enemies and taken over the habitats of millions of other species in what is now called the Holocene, the age of humans. We are also responsible for the sixth mass extinction event, where species are disappearing at a rate that hasn't been seen since the dinosaurs bid farewell to the planet.

Throughout this remarkable phase of our evolution, which has taken place in the last two centuries, we enjoyed economic growth and the prosperity that came with it.

Then came the fog.

December 5, 1952, began the deadliest case of air pollution to ever smother a city. For five days, London, England, was choked by a toxic fog that claimed twelve thousand lives. London was the first industrialized city in the world and famous for its fog, which was partially a result of its maritime climate but also because coal was burned openly in homes and factories. Black smoke particles hanging in the air provided nuclei, or seeds, around which water vapour condensed into droplets, intensifying the fog. For almost a week, a temperature inversion held cold stagnant air close to the ground over the city, so smoke did not rise. Visibility was reduced to a few metres even during the day. Buses had to be guided by conductors on foot carrying lanterns. At the same time, chemical reactions between sulphur dioxide in the smoke and water droplets in the air created a toxic brew of sulphuric acid that burned the lungs, resulting in one hundred thousand people suffering from respiratory illness.

This was the first time the environmental impact of fossil fuels literally hit us in the face and when we began to do something about it.

Following that environmental disaster, the British government enacted the world's first clean air act in 1956. It restricted open burning of coal and provided subsidies for cleaner, more efficient technologies to burn smokeless fuel, and the skies over London eventually cleared. It was the first time that a federal government took positive action to rectify an environmental problem caused by burning fossil fuels. And it worked.

When industrialization came to North America, cities such as Los Angeles became choked in a brown haze caused by emissions from vehicles, a combination of smoke and fog that became known as smog. By the 1970s it was difficult to see the city skyline as chemicals coming from millions of tailpipes, with a little help from sunlight and moisture, created toxic brown particles that burned eyes and irritated lungs. You could literally taste the air.

Smog was so visible and posed such a threat to health, California introduced strict emission standards for vehicles in the 1970s, including the use of catalytic converters on exhaust systems to catch those harmful chemicals. While there were protests from the auto industry, claiming the rare metals such as platinum needed for the converters would make cars too expensive, the laws held and the skies cleared. A new industry was also created recycling catalytic converters.

A more recent graphic example of the effect fossil fuels have on the atmosphere was the clear skies that appeared over cities such as Beijing, China, and Seoul, South Korea, during the COVID-19 outbreak of 2020. Since people were locked down to prevent spread of the disease, industries shut down and vehicles disappeared from highways. Skies shrouded in smog for decades because of extensive use of coal and diesel appeared blue once again. It was a visual reminder not only of our direct impact on the environment but also of how quickly nature can respond when we stop pushing it. But clearing the skies is only half of the problem with burning fossil fuels.

A greater issue is what we don't see coming out of pipes and stacks, the invisible hazard seen mostly by science: the carbon dioxide and other greenhouse gases that have been changing the climate. The issue has been known for decades; in fact, the very first documentary I wrote for CBC Radio's *Ideas* program in 1977 was on climate change. It was part of a series called "Running Hot and Cold," because at that time, the scientific community was somewhat divided as to which way the climate of the Earth would go in the future. On one side, geologists argued that there have been five ice ages in the past with so-called interglacial warm periods between them, each lasting roughly ten thousand years. We have been living in one of these warm periods, but it has been twelve thousand years since the last ice age, so we are overdue for another cold snap.

On the other side of the argument were the climate scientists studying the greenhouse effect, who argued that the warming effect of these gases introduced since the beginning of the Industrial Revolution would overpower the natural ice age cycle and force the climate to warm up. That was more than forty years ago. Now, as we watch global temperatures rise at an alarming rate; heat waves, wildfires, and droughts increase in frequency; powerful hurricanes and cyclones become more common; ice vanish from glaciers in the Arctic; and many other visible effects of a warming climate, we know who won that debate.

But action to reduce those emissions and switch over to clean alternatives has been slow, thanks to heavy lobbying from the fossil fuel industry and political fear of harming the economy. Despite big announcements at world summits such as Rio (1992), Kyoto (1997), Copenhagen (2009), Paris (2015), and Glasgow (2021), emissions have continued to rise steadily. Those clear skies over cities during COVID-19 are an example of the kind of reduction in fossil fuel burning we will need every year to meet the international goal of preventing the Earth's climate from warming beyond a further 2°C in average temperature.

The COVID pandemic was also a dramatic demonstration of how people worldwide can instantly change behaviours in the face of a crisis. First, science pointed out a new threat to humanity: a deadly virus that spreads through the air. Hospitals were filling rapidly with new cases that were appearing around the world. We were told we had to "flatten the curve" of rising cases before they overpowered our health care system. The governments, on the advice of science, introduced mandatory lockdowns and guidelines for wearing masks, handwashing, and social distancing. Then billions of dollars were spent supporting research into vaccine development and providing for industries that lost income and for those who lost their jobs. Pharmaceutical companies ramped up production of the vaccines and emergency clinics were established to inoculate the population. Those were drastic moves, but everyone (well, mostly everyone) from the top down bought into it and the curve was flattened more than once. That is the scale of change that is required if we are to flatten the curve of rising carbon emissions.

Stepping away from fossil fuels will not be easy. The industry is huge, with fingers into so many areas of our lives. Shutting it down completely overnight would cripple the world economy and cost millions of jobs, not just in the oil industry itself but also in the transportation, chemical, and plastics sectors as well as those related to many other products and services derived from oil. This has created a conundrum for governments caught between the need to preserve jobs and the economy by supporting industry, while upholding international commitments such as the Kyoto Accord and the Paris Agreement to cut emissions and protect the environment.

Sadly, there is no magic bullet, no single clean alternative that can be slotted in to replace fossil fuels and provide us with an equivalent amount of clean energy packaged in such a dense form. It comes down to three fundamental issues: energy density, energy storage, and energy efficiency. The only fuel denser than fossil fuel is nuclear, where a few kilograms of uranium contain the same amount of energy as a tonne of coal. But nuclear power comes with a unique set of issues, technical, social, and political. However, it is emissions free and provides power 24/7, so it is being reconsidered in a newer modular form.

A huge challenge is how to carry around other forms of energy in a container like we do with a tank of gasoline. Batteries are very heavy, can barely hold as much electrical energy as an equivalent tank of gas, take a long time to recharge, and are made of expensive rare-earth metals. A great deal of effort is going into improving batteries and developing other ways of storing energy for future use, but many potential solutions are still in the research phase. In the meantime, we use electricity as we make it, which means the current running through the lights in your house was generated just a few minutes ago.

Of course, you can't mention energy production without including the issue of hazardous waste. Interestingly, since nuclear waste is highly toxic with a lifetime of thousands of years, all of it is stowed under lock and key. Every other source of energy production throws its waste into the environment. Solar panels, which are not recyclable at the moment, will end up in the waste stream when they come to the end of their thirty- or forty-year life cycles. Windmills kill birds and bats; hydroelectric dams flood landscapes and block migrating fish. On the other hand, the World Health Organization estimates that between seven and nine million people die every year due to air pollution from burning fossil fuels.[8] No matter what form of energy production you choose, fossil fuel or otherwise, there is always a cost that must be factored into the decision of which is best to use.

Even the manufacture of clean energy sources involves carbon emissions. It takes high temperatures to make solar cells, energy to manufacture and transport wind turbine components to their operating sites, and mining of elements such as lithium to produce batteries—all require burning of fossil fuels. Not until these clean alternatives dominate the energy sector will it be possible to make solar cells from energy produced by solar cells and for windmills to make more windmills.

This is not the first time we have been faced with the challenge to find other sources of energy beyond fossil fuels. When the first oil crisis hit North America in the early 1970s, there was fear that the world oil supply would run out, which led to intense research and development of clean alternatives. So the fundamental work into those technologies has already been done. In other words, much of the clean energy technology already exists. It just needs to be employed on a larger scale. The new incentive of climate change has science once again looking at clean alternatives but with renewed vigour. The challenge is how to fill the gigantic energy hole that would be left behind when fossil fuels are removed from the picture.

Thankfully, there is no shortage of clean energy on Earth.

Feel the power of the sun burning your skin on a hot summer day. Look at the force of the wind whipped up by hurricanes that destroy entire cities. Watch the energy within the Earth erupt—in explosive volcanoes such as Mt. Vesuvius that destroyed the city of Pompeii or in tsunamis like the one in Thailand that wiped out towns and villages—as its crust constantly shifts in the process of plate tectonics. Every year towering flames from wildfires scorch the land as all that biomass goes up in smoke. These forms of natural energy surround us and hold tremendous power. All we have to do is learn how to harness them in more effective ways than we have in the past.

For centuries we have captured the free energy of the wind to sail across oceans and explore the planet. Windmills have been pumping water and generating electricity for more than a century. Ancient

Greeks captured sunlight with curved mirrors to start fires, now silicon panels convert sunlight directly into electricity. Heat pumps utilize geothermal energy from the ground. We are awash in energy.

The challenge with capturing wind or solar energy is that while they are essentially free, they are spread out over large areas. To generate the equivalent amount of energy contained in a tonne of coal or a barrel of oil takes many towering wind turbines or solar arrays covering a lot of land. The energy density is low. Then there is that little problem where the wind doesn't always blow and the sun doesn't always shine. So new technologies are being developed to store that energy for when it is needed.

The transition towards clean energy sources and efficient means to use that energy will not happen overnight. It will be an evolution rather than a revolution. Revolutions are usually violent events, where two opposing sides both dig in their heels, leading to conflict, people getting hurt, and often chaos in the aftermath. Pitting tree huggers against oil magnates is not productive.

Evolution is taking existing technology through incremental steps, until it grows into something better. The principle behind combustion engines in cars has been around for more than 150 years. It has served society well. And just as cleaner and more efficient gas- and diesel-powered engines replaced coal-fired steam engines of the past, the time has come for the next phase of engine evolution. Electric motors are cleaner and more efficient than combustion, so it is an easy swap under the car hood. But then there is the issue of where the electricity comes from and how to store it. These issues are not insurmountable. We have seen the evolution of technology before. Think of how the telephone has changed: from a tabletop device patented by Alexander Graham Bell in 1876 as a means of transmitting voice only over wires to the astounding devices we carry in our pockets with too many functions to mention. Bell would turn in his grave several times if he saw a smartphone. That is evolution of technology.

If we can do it for communication, can we do it for energy?

The answer is yes.

We are reaching a turning point where more countries are moving away from fossil fuels towards electrification. And the technology to generate and use electricity has been around a long time.

In 1831, British scientist Michael Faraday's discovery of the relationship between magnets and electricity made him famous. He found that passing a magnet through a coil of wire produced an electric current in the wire. That fundamental concept is behind electric generators and motors that power our civilization today. Generating stations use huge coils of wire surrounding powerful magnets that are mounted on rotors and spun rapidly to generate electricity. The challenge has been how to spin the magnet.

Some generators are spun by the force of falling water, through natural hills such as Niagara Falls or artificial hills of water behind hydroelectric dams. More commonly, generators are driven by steam turbines, which are basically a series of windmills contained in a tube. When high pressure steam is run through the tube, the blades spin, which turns a shaft attached to the generator. The question then is where to get the heat to boil the water to make the steam.

Coal, oil, or natural gas have been convenient fuels to burn in boilers because they have been cheap and abundant. Nuclear reactors, which are basically hot water kettles (known as the most complicated way to boil water), use the heat of nuclear fission to boil water into steam then run it through turbines. Even nuclear fusion, which duplicates reactions happening at the centre of the sun involving temperatures above a hundred million degrees, still just boils water to make steam. And it doesn't have to be steam. Any gas under pressure can spin a turbine, even just compressed air.

But there are other ways to spin the magnet in a generator. The massive blades of a wind turbine are connected directly to a generator, while underwater tidal energy does the same thing in a marine environment. You can drop a weight from a great height on a pulley so that as the pulley's wheel spins it runs a generator.

Here's the fun part. If you run a generator backwards, that is, put electricity into it, the generator becomes a motor that can drive wheels. That's why electric vehicles have motors that take electricity from the batteries to propel the car forward, but when the driver lets up on the accelerator, the motor becomes a generator turning the movement of the vehicle as it goes downhill or comes to a stop back into electricity that is returned to the battery. Both functions are performed by the same device. Our challenge is to find innovative ways to spin magnets and efficient ways to use that spin.

This book is a look at what a clean green future could actually look like: the vehicles of the future, energy sources, and lifestyles that can realistically enable society to achieve a sustainable lifestyle without crippling the economy, or sending everyone back to the caves and trees.

Future energy will come from many different sources, and less from large, centralized generating stations. Alternative energy sources such as wind, solar, and geothermal have been available for decades. But they alone will not be enough. Additional power will come from small nuclear reactors that are buried underground to power small towns. Energy will be captured from waves, tides, hydrogen, and fusion. Your own home will become an energy producer, with windows, walls, even your own clothing having the ability to produce electricity. And finally, there are the innovative sources, today's science fiction if you will, such as space-based solar power satellites with enormous mirrors to capture sunlight that's converted it to microwaves beamed to the ground to light up entire cities.

Along with this new technology will be a renewed respect for energy use and a clean environment. Cities will adapt to more public transit and village lifestyle, where walking replaces wheels. Vehicles will no longer have tailpipes that emit smog particles.

Climate change is now in our face, no longer something for future generations to worry about or in far-off places such as the Arctic where polar bears are affected. It is here now, a clear and present danger with record-high temperatures year after year, decade after decade, large areas of land burning up, water supplies dwindling. The time has come to move forward in realistic ways.

Green technology such as wind and solar are already less costly than fossil fuel with other forms following suit. It is also one of the fastest developing sectors of the economy, with tremendous potential for growth as current products improve their performance and new products come online. That means jobs, and more jobs, and a stronger economy.

A new green age is upon us.

Archimedes using a mirror to reflect sunlight onto Roman Ships during the battle of Syracuse.

Chapter One

Solar Power

*The most abundant form of clean energy. How much
land, or rooftop area, is needed to meet our demands?*

You are a Roman centurion aboard a naval battleship approaching the city of Syracuse on the island of Sicily. Suddenly, a brilliant flash appears on the shoreline, blinding you with light as bright as the sun. You feel tremendous heat searing your skin, and as you fall to the deck and look up, the sails above you burst into flames.

Such is the legend of a battle during the third century BCE when a clever inventor attempted to use the sun as a weapon of war. The Second Punic War was raging between the empires of Carthage and Rome over territory around the Mediterranean. The Carthaginians, who occupied much of what is now Spain and parts of North Africa, were seeking to conquer Italy, but the Romans were a formidable force, fighting back with a superior naval fleet. With battles at sea proving difficult, the Carthaginians made a brilliant and unexpected move by attacking Italy on land. Under their supreme strategist and general, Hannibal, an army of ninety thousand troops (estimated number varies greatly, which often happens when legendary leaders are involved) and fifty elephants marched a thousand kilometres from Spain across Europe to the Alps, then crossed the rugged mountains and attacked Italy from the north.

Many lives and elephants were lost along the arduous journey, but as cities were defeated, more men were recruited into Hannibal's army, keeping it strong and invincible. Avoiding the city of Rome itself, the army moved south; as cities fell, they switched their allegiance from Rome to Carthage. A prime goal was to reach the island of Sicily, which lies just off the toe of the "boot" of Italy. The fertile soils of the island provided an important source of grain and other food products for both Italy and Carthage.

Syracuse, along the southeast coast of the island, was a key city state. It was an important cultural and trading centre where the local king had sided with Carthage.

The Romans wanted it back.

A fleet of warships was sent to the island, but they would face a unique challenge thanks to a very special resident who called Syracuse his hometown: the brilliant mathematician and engineer Archimedes. Famous for his "Eureka" moment when he figured out how to measure the gold content of a crown by immersing it in water to calculate its density, he was also an inventor whose engineering experiments led to the creation of some innovative devices.

As the heavily armed Roman ships attacked the harbour of Syracuse, they were met by an unusual defence: bizarre machines never seen before in battle lined the shore. Built under the direction of Archimedes, the ingenious war machines were onagers, a variation on the catapult that could hurl heavy objects great distances with incredible accuracy. Similar weapons had been used before, but one device, called Archimedes's claw, was completely new. It used an enormous lever arm with a grappling mechanism on the end to hook a ship and lift it up on its end or capsize it on the spot, spilling sailors into the sea.

But the most legendary weapon developed by Archimedes was one designed to set fire to Roman ships. There is debate about whether this was actually done or whether it was just a legend that grew out of respect for one of the greatest mathematical minds of the ancient world. The story goes that Archimedes was able to concentrate sunlight into a powerful beam that was then aimed at enemy ships. One idea is that he used a very large parabolic mirror, several metres across, precisely shaped and steerable, to create the beam. Smaller parabolic mirrors were used at the time by the Romans and Greeks to start fires, a tradition that is still used today to light the Olympic flame. Manufacturing one large enough to burn a ship, or at least set its sails on fire, would have been an engineering challenge in those times, but not impossible.

However, as legend has it, Archimedes may have accomplished the same goal by reflecting sunlight off the shiny bronze shields of soldiers lining the shoreline. Imagine hundreds of soldiers forming a curved line along the shore, all angling their shields towards the sun to form one large parabolic reflecting surface. If positioned correctly, the light could be focused onto the ships and create enough heat to ignite the sails and wooden decks.

That's the legend. The Roman sailors were reluctant to approach the harbour, but eventually they did return and take the city at night. Archimedes was later killed there by a Roman soldier who didn't realize who he was.

Whether or not Archimedes did set ships ablaze, his fundamental concept that sunlight carries energy and can be converted into heat was correct. Today, it is one of the fundamental approaches to solar energy, where mirrors are used to concentrate sunlight to heat a fluid and generate electricity. It just takes a lot of mirrors.

Sunlight is the ultimate energy provider for planet Earth. It is the most abundant natural energy source. Also powering the wind and weather, it drives mighty ocean currents and nurtures life. The amount of energy in sunlight that falls on 1 square metre of the Earth's surface is roughly 1 kilowatt (kW) per second, or 3,600 kW per hour (kWh). An average American home uses about 1,000 kWh per month, which is a tiny fraction of the sun power that hits its roof. Every year, the sun pours 200,000 terawatts (TW) of free energy down on the planet. A terawatt is 1 trillion watts. A light bulb takes 60 watts. Our global energy demand is small by comparison, only 20 TW, which

Solar thermal plant with many mirrors reflecting sun onto a central tower.

© Samuel Faber Pixabay

means sunlight could cover all our energy needs a thousand times over. Another way to put it is that the amount of sunlight falling on the Earth in one hour could cover all our energy needs for a year. In fact, solar energy is so powerful we sometimes run from it with sunshades and sunscreen. It can literally burn us alive. All we have to do is find innovative ways to gather it up and put it to use.

So how much land would be needed to power civilization on the sun alone?

Dr. María Bernechea Navarro, senior researcher, ARAID Foundation, Instituto de Nanociencia y Materiales de Aragón (INMA), CSIC-Universidad de Zaragoza in Spain, states: "According to a study, six points on the Earth each at 100 square kilometres covered with solar panels of an efficiency of 10 percent, could provide enough energy for the world."[1] In other words, with current technology, the entire world could meet all its electricity needs on solar power alone using an area of 500,000 square kilometres, or 200,000 square miles. That is roughly the size of Spain. The United States could be powered with a single solar farm the size of Kansas, and in Canada that's slightly more than New Brunswick, PEI, and Nova Scotia combined.

Fortunately for the residents of Spain, Kansas, or the Maritime provinces, centralizing solar power production into a few huge projects would not really work. That energy would have to be sent vast distances to the rest of the world along power lines, which would involve too much energy loss along the way to make it worthwhile. Thankfully, sunlight shines over huge areas of the Earth so there are many places available to gather it up.

Solar power is produced in two fundamental ways: concentrated thermal, where a circular array of mirrors focuses sunlight onto one spot (what Archimedes was trying to do) to produce heat to make steam and run turbine generators, and photovoltaic, where special materials convert sunlight directly into electricity with no moving parts.

Archimedes was trying to take advantage of the power in sunlight with his shiny shields, but that power is somewhat diluted. Experts agree that it is unlikely that even a large group of soldiers all aiming their shields on one spot could have collected enough sunlight to burn an entire ship.

Several modern groups have tried to duplicate the Archimedes legend, including the popular science TV show *MythBusters*. In 2010, they gathered five hundred students from MIT who held mirrors along a shoreline and reflected sunlight onto a small wooden sailing vessel. The objective was to set the ship's sail on fire, which was even painted with a large black mark to help absorb the solar energy. While they did manage to raise the temperature on the sail to an impressive 116°C, it was not nearly high enough to set it ablaze. However, one of the hosts of the show, Jamie Hyneman, who was aboard the ship during the experiment commented that having brilliant sunlight focused on the deck from so many sources was extremely distracting and would have made it difficult for soldiers in battle to see their target. Perhaps that was Archimedes's strategy.

Other experiments using multiple mirrors mounted on arrays have managed to create a little smoke on the side of a wooden boat, but the boat had to remain stationary for a long time, which seldom happens in battle. It took a while to bring the wood up to temperature, and the smouldering was so slight that a single crew member could easily douse it with a bucket of water.

Fortunately, the idea of turning reflected sunlight into heat did not die with Archimedes. His fundamental concept is still used today on a much larger scale in places such as the deserts of California and Spain. Gigantic solar arrays involving thousands of moveable mirrors arranged in a circular pattern focus sunlight onto a central tower. Inside the tower is molten salt, which can reach ultra-high temperatures above 550°C. Heat from the molten salt is used to make steam that runs steam turbines which spin generators to make electricity. These large installations, covering many square kilometres, produce amounts of electricity similar to that of conventional generating stations. But in this case, the energy source is provided by clean, free sunlight rather than fossil fuels or the heat from nuclear fission.

However, there are limitations to the concentrated thermal system because of losses during three energy transformations. First, sunlight is converted into heat energy, which is then turned into the mechanical energy of the spinning turbines, which is finally converted into electrical energy. In the end, only about 20 to 40 percent of the energy in the sunlight becomes electricity.[2] On the other hand, conventional photovoltaic cells, the type seen on rooftops, can achieve about 30 percent efficiency with no moving parts at all.[3] One criticism of solar energy is that it only works when the sun is shining. The big advantage to the thermal system is that the molten salt in the central tower is very good at absorbing heat and is able to store that heat for many hours. The plant can continue to produce electricity even after the sun goes down.

Concentrated solar power is best suited to hot dry regions with clear skies, generally between 15 and 40 degrees latitude from the equator. That includes the Middle East, the deserts of southern Africa, India, China, Australia, and the American Southwest. All these regions are currently developing or planning solar megaprojects. Archimedes would be impressed.

One variation on the solar thermal principle involves long trough-shaped mirrors that focus sunlight onto black tubes filled with oil that run down the centre of the trough and are heated to high temperatures. The hot oil is used to turn water to steam. Some operations use what are basically magnifying lenses to concentrate the sunlight; others employ sterling engines, which are mechanical devices placed in front of parabolic mirrors, and use the heat to run electric generators.

The easiest way to convert sunlight into energy is to use it to provide hot water. In sunnier regions, relatively simple rooftop devices are made with pipes painted black and covered with glass. The black absorbs sunlight and gets hot, like pavement in summertime. Water is circulated through the pipes, heated by the sun, and then stored in a hot water tank, providing all the hot water needs of a house. Even sailors use sunlight to take hot showers on the deck of a boat. A device called the solar shower is just a black plastic bag that is filled with water and hung from the rigging. A valve at the bottom lets the water shower down on a salty sailor. The bag comes with a warning that a person can be scalded if the bag is left in the sun too long—that's how powerful sunlight is.

Both systems, concentrated solar using mirrors and photovoltaic using solar panels, have advantages, but when it comes to very large-scale deployment around the world, photovoltaic has become the more popular choice, in part due to cost.

Photovoltaics came down in cost much more rapidly than concentrated solar power. Market development in countries like Germany triggered upscaling of production and installation, which was very effective in driving cost down because photovoltaics is modular and highly standardized. This is illustrated by the well-known price experience or learning curve. Concentrated solar power is much less modular, not yet strongly standardized, and has not benefited strongly from upscaling. The main advantage to concentrated solar power is that, in principle, heat can be stored relatively easily so you can either stretch the length of the day or even have 24/7 power production. However, with the rapid development of electricity storage in batteries and fuel production with electrolysis, this advantage is temporary. The cost per kilowatt hour of photovoltaics is probably unbeatable by any other alternative.

We are rapidly approaching the one cent per kilowatt hour in sunny regions, which is an order of magnitude lower than where concentrated solar power is today.

Wim Sinke, professor of photovoltaic energy conversion, TNO—Netherlands Organisation for Applied Scientific Research

Photovoltaics, or the photoelectric effect, is a much simpler system to operate because there are no fluids to pump through tubes, no super-hot liquids to handle, no turbine generators. It is currently the cheapest form of solar energy because of the dramatic drop in price of solar panel manufacture. Research is looking to find new materials that produce electricity directly when sunlight shines on them.

The idea that sunlight can knock electrons free in certain materials and produce a small electric current was first discovered in 1839 by French physicist Alexandre Edmond Becquerel. He and others who followed worked with minerals such as selenium and platinum, but the electric currents produced were very small—capturing only a few percent of the energy in the light—plus, the materials were expensive, so photovoltaics was an interesting curiosity, but not yet practical. A few years later, Charles Fritts developed the first selenium solar array on a rooftop in New York City, showing that current was produced even when the sun was partially covered by cloud. He believed at the time that solar energy would eventually replace coal.

What Fritts and other experimenters did not know was exactly how sunlight produced the electricity. They suspected it had something to do with heat, but it was Albert Einstein, using the quantum theory of light, who figured out in 1904 how the photoelectric effect works. That discovery (not his theory of relativity for which he is most famous) won him the Nobel Prize. He described light as a stream of tiny packets of energy called photons, or light quanta, that are absorbed by certain materials. When a photon strikes an atom, it gives the electrons around the outside of the atom extra energy. If that energy is greater than the force that is attracting the electron to the nucleus of the atom, the electron will fly off carrying that absorbed energy with it. It's like a stone dropped into water. The tiny drops that fly upwards in the splash are carrying the energy of the stone. The amount of energy the electrons carry depends on the type of material and the frequency, or colour, of light that hits it.

Basically, it is light energy coming in and free electrons flowing out. Those free electrons are the electric current that we can put to work. So, the trick to developing efficient solar cells is to find the right materials that will freely give up their electrons when absorbing light.

The Nobel committee recognized that Einstein's discovery had great potential in many areas of science, but early solar cells were still very inefficient. They converted only a small percent of the light that fell on them into electricity.

It was an accidental discovery in the 1950s at the Bell Laboratories in the United States that gave solar power the boost it needed to be taken seriously.[4] Silicon, which was being used to develop the first electronic computer chips, was found to also have a photoelectric effect, with a tenfold greater efficiency than selenium. Silicon is one of the most common elements on Earth, found in everything from rocks to beach sand, so supply is not an issue. However, silicon is not the only ingredient in a solar cell. Other elements, such as phosphorous, are added to increase the number of available electrons.

At a microscopic level, the silicon molecule looks like little cubes. When heated to a high temperature and melted, these cubes line up beside each other in a uniform sheet as one giant crystal of repeating units. Electrons move more freely through crystals, so the silicon is made into large sheets that can be cut into wafers for computers or left large for solar panels.

The principle sounds simple, but making silicon co-operate in such an orderly manner is an exacting science. According to Dr. María Bernechea Navarro:

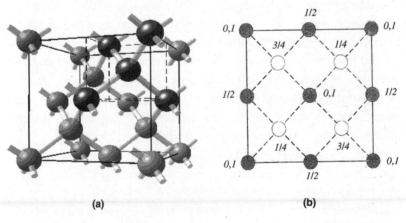

(a)   (b)

The cube-shaped structure of a silicon crystal.

Silicon must be super pure to be used in solar cells. It loves oxygen so you can't leave it around because it will quickly become silicon oxide. That is why it must be encapsulated or protected.

The silicon crystal must be more than 99.9 percent pure for the photoelectric effect to work. This is hard to do on a large scale, so the first solar cells were prohibitively expensive, costing more than $100 per watt of electricity they produced.[5] As is often the case with innovative new technology, it was the military, with its deep financial pockets, that funded research on solar cells for their own purposes: to keep satellites operating in space for long periods of time so they could spy on enemies from above.

How a Solar Panel Works

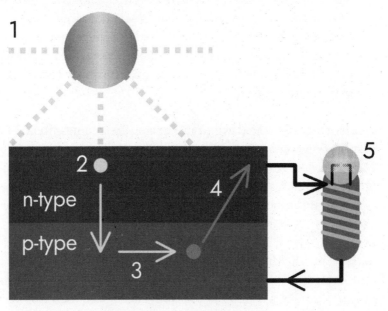

Cross-section of a solar panel showing flow of electrons.

Image courtesy of Chris Woodford/explainthatstuff.com

A solar cell is the small device that converts sunlight into electricity. Cells can be joined together to form solar modules, which can be assembled into even larger units of solar panels.

This modular form allows solar panels to be as small or large as you like, powering hand-held devices, covering the roof of a house, or blanketing a farmer's field.

A solar cell has two layers of silicon with a thin barrier between. The top layer is infused with tiny amounts of an element such as phosphorous, which has an excess of electrons. The bottom layer may be infused with boron, which lacks electrons. These whole atoms cannot cross the barrier on their own. To make the electric current, sunlight penetrates to the bottom layer and knocks electrons off, which then have enough energy to cross the barrier and pop up to the top layer, giving it a negative charge. The atoms left behind in the bottom layer lack electrons, giving that layer a positive charge. Wires at the end of the panels connect the two layers, and the electrons flow through the wires as a current from the top layer back down to the bottom layer. As long as sunlight is pumping energy to the atoms, the electric current continues to flow.

Scientists and engineers are constantly working to improve the process with different elements to get as much current as possible from the sunlight.

The 1950s saw the dawn of the Space Age and heightened Cold War tensions between the United States and the Soviet Union. Both countries were competing to develop bigger rockets as well as more powerful satellites to look down on each other from the high ground of space. But a satellite orbiting the Earth will only pass over an enemy target once every hour and a half, so it must make many orbits over days, weeks, and even years to spot any developments such as building missile launch facilities on the ground. Batteries do not last long enough to keep the eyes in the sky running for long periods; however, sunlight is four times stronger in space than it is on the ground. If a satellite is placed in the right orbit, the intense sunlight can provide almost continuous power.

Companies such as AT&T and Bell Labs began working on the chemistry and manufacturing to improve efficiency in solar cells. It enabled the Americans to launch Vanguard 1 in 1958, the first satellite powered by solar energy. It operated for eight years. The Russians powered Sputnik III with solar cells, as did Canada, the third country in space in 1962, with its satellite Alouette 1. Since then, virtually all satellites carry solar panels. The International Space Station is also entirely solar powered, with arrays extending out to an area larger than a football field. But these exotic cells are made of multiple layers of expensive materials such as gallium arsenide, germanium, and other elements that are best suited to military budgets.

Solar cells commonly used today have dropped 99 percent in price since the 1970s. Much of that cost reduction has come in the last two decades, largely because of research and development done in China.

> Scaling up, primarily done by the Chinese, was very successful. If scaling up had been done elsewhere in the world, it also would have worked. What the Chinese also did very well was build up the entire supply chain from silicon to solar modules and even complete systems. That is different from producing feedstock in one country, cells in another country, and modules in a third. That has added to the effect of scale and that is why the price reduction, or learning curve, has become steeper than it already was over the past ten or fifteen years.
> *Wim Sinke*

Solar cells average around 18 to 22 percent efficiency, which really means that about 80 percent of the sunlight falling on them is not being used. That low efficiency is the biggest drawback to solar cells (besides the fact that the sun isn't always shining), which means it takes a lot of them spread over a wide area to produce enough electrical power to be useful. That's why solar farms are so huge. Improving that efficiency would go a long way towards generating the same amount of power out of a much smaller area—say, the roof of a house, for example. In 1961, scientists William Shockley and Hans Queisser calculated that solar cells can never exceed 30 percent efficiency, the so-called Shockley–Queisser limit.[6] This is because some of the sunlight is turned into waste heat, some is reflected off the surface, and not all wavelengths, or colours, of light are absorbed. However, laboratories around the world are working to exceed that theoretical

Perovskite thin solar cell.

© Dennis Schroeder / National Renewable Energy Laboratory

limit by stacking the cells into multiple layers or adding thin films of other photovoltaic materials on top. Thankfully, this research is showing results that exceed that limit.

One promising advance in photovoltaics is a family of materials called perovskites, which have the cube-shaped structure like silicon but are easier to manufacture. The first mineral of this type was discovered in Russia's Ural Mountains in 1839 by Lev Perovski, after whom the metallic-looking mineral is named. In 2009, Japanese researchers found that perovskites could absorb photons of sunlight and turn them into electricity, and were easier to work with than silicon. They can be made at room temperature and, unlike the rigid, brittle sheets of silicon, can be mixed into a solution and sprayed into thin sheets only a few molecules thick using inkjet printer technology.

Perovskites can be printed onto flexible sheets of plastic or almost any material and, in laboratory experiments, show an efficiency of 25 percent, up from just 3.2 percent since 2009. That is the most dramatic rise of any solar technology. Silicon took decades to achieve that.

Perovskites hold great hope because of their lower cost and because they can be manufactured at room temperature. However, they cannot completely take over from silicon. They still need to be proven economically feasible and durable on a large scale. It is more likely that they will augment silicon by providing an extra layer to boost power, or be used in areas that silicon cannot reach. One promising avenue is layering a thin film of perovskite on top of silicon, so their combined efficiency exceeds 30 percent.[7]

Rather than replace silicon solar cells, which will still be used in large-scale commercial applications, perovskites will find their way into many places not previously considered for gathering solar energy. Since the discovery of perovskites, laboratories are looking at different formulas for their development, including the use of nanoparticles.

> Nanoparticle semi-conductors . . . they adapt the range of wavelengths they can absorb.
>
> Not just for solar cells, because they are in solution, they can be used in a variety of situations, paints, clothing, because they can be printed.
> *Dr. María Bernechea Navarro*

When you look at city buildings, the rooftops seem to be the obvious place to catch the sun. But in a tall building, the roof is the smallest part of the total exterior space. The sides of the buildings add up to a much larger area, so scientists envision using that expanse to capture sunlight through perovskite solar coatings, either

When seen from above, the rooftops of city skyscrapers are small compared to the surface area of the glass-covered sides of the buildings.

in the form of paints or incorporated into building materials. Experimenters are working with solar window blinds and awnings that generate electricity while blocking out the sun. Perovskites can also be woven into flexible fabrics such as tents or, perhaps one day, even clothing. Imagine putting your device in a pocket and it is charged by your shirt.

One drawback to perovskites is the small presence of lead in their composition, which raises concerns about it leaking into the environment. There are also issues with perovskites being sensitive to moisture and temperature, so research continues to find materials that are cheap, safe, and efficient. Many new solar materials are very efficient but expensive; others, such as organic dyes made from plants, are cheap but their efficiency and stability is low. While perovskites are performing well in laboratory experiments, it will take some time before the perfect combination of materials and manufacture on a large scale will become commercially successful.

An interesting variation are solar cells that are transparent and look like window glass. That seems contradictory because solar cells are supposed to absorb as much light as possible not let light pass right through. One group at UCLA and UC Santa Barbara in California used tiny silver nanowires coated with titanium dioxide. The microscopic particles are photovoltaic and, when embedded in a sheet of plastic, produce electricity while still allowing 70 percent of the light to pass through. That means they could be used as window-tinting material, giving off a pale-blue colour.

> You can apply perovskites as a regular thin film, either on glass or flexible plastics. But by making small holes, or stripes, in the thin film, you can let part of the visible light go through, so you can make a windowpane that lets white light through. Of course, the intensity would be less. You can also make the cell so thin that the cell itself becomes transparent, but then it will not be white light that comes through, but it will be coloured light, which also has many applications.
> *Wim Sinke*

Another team at the University of Michigan developed a material that is completely transparent, at least to the human eye. Their secret is panels that do not need every wavelength of light that falls on them. We have all seen a spectrum, the multiple colours of a rainbow that make up visible light. Each colour is a different wavelength, with blue light made of shorter waves and red light longer. But there are other wavelengths of light that we don't see, such as very short ultraviolet and extra-long infrared. The transparent cells take advantage of these. By adding special salts to the material, infrared light is absorbed, then transmitted to the edges of the glass where solar voltaic materials tuned to that wavelength of light turn it into electricity. Visible light, which is not used, passes right through undisturbed, so the material can remain transparent without affecting energy production.

> Perovskites and nanotechnologies will coexist. Silicon for large scale, thin film clothing, windows and smaller applications, including indoor . . . Solar can be integrated almost anywhere—textiles, windows, indoors, the options are there—but whether we do it is up to us.
> *Dr. María Bernechea Navarro*

That's right—solar power generated indoors. Think of solar coatings in your home as recycling light. We shine lights everywhere but not all of it is used. A lot is absorbed by walls and turned into heat, or it escapes outside. If you look at pictures of the Earth at night taken by astronauts aboard the International Space Station, cities shine like galaxies in the dark, which is beautiful, but all that light beaming upwards is light we are throwing away. Even capturing a small percentage of that with indoor solar coatings would save a lot.

While the efficiency of transparent solar cells is only a small percent at the moment, when you think about the area of all the windows in office buildings, high-rise condos, and apartment complexes, it adds up. Future designers can incorporate solar surfaces into the architecture of the buildings so that you don't even know they are there, adding function to beauty in the design. On a smaller scale, these materials could replace the transparent faces of smartphones and devices to help batteries last longer. Size doesn't matter: all these ideas contribute towards increasing the area outfitted with solar panels without using up any more valuable land.

Even if we were to cover all the buildings and windows in a city with solar voltaics, the electricity generated would not be enough to meet our voracious energy appetite. There will still be a need for large solar farms, which cost less and can be built more quickly than conventional generating stations.

But questions immediately follow: Where to build them? Will they take up land that could be used for agriculture, or space that is environmentally sensitive?

There are a number of approaches to deal with this issue.

Before the installation of the giant Stateline Solar Farm in California, located in a large dry lakebed called the Ivanpah Valley, an extensive environmental study was conducted to assess the impact of the solar farm on the desert region and tortoise in particular. It was determined that the solar projects would have a negative effect on the desert habitat. Fortunately, the tortoises are found over a huge region covering California, Nevada, and Utah, and the solar farms would cover less than 1.5 percent of the total region.[8]

To protect the tortoises, conservationists combed the proposed construction site and relocated hundreds of the animals to surrounding similar habitats. Care was also taken to ensure that connecting pathways allowed the tortoises to move throughout the region. An even better idea is to make eco-friendly, or even eco-positive, solar parks. This kind of co-operation between conservationists and energy developers is a model for effective management of an ecosystem while new energy projects move forward.

A criticism of solar energy is that it will steal sunlight needed to feed green plants that rely on photosynthesis. One way around that is through a new concept: agrivoltaics, which allows plants and solar panels to coexist. This is particularly useful in hot regions with limited rainfall. By spacing the panels apart, sunlight can penetrate to the plants growing between. The shade provided by the panels lowers the temperature of the ground, reducing heat stress and enabling the ground to hold moisture longer. Researchers are also looking into incorporating photovoltaics into the glass of greenhouses to see how that affects plant growth: it would convert only infrared light (heat) into electricity while allowing other wavelengths to pass through. In line with that co-operative spirit, electricity generated by solar panels can be used by the farm itself or sold to the grid, so the farmers can be harvesters of both food and sunlight.

> Perovskites have a very specific cut-off, or wavelength of light, so that if you put it on top of something else, it absorbs strongly in one part of the spectrum, but lets the other part of the spectrum through almost perfectly. That is what you want to build a high efficiency tandem, and that is very difficult to achieve with other types of low-cost semiconductors.
> *Wim Sinke*

Experiments with multi-junction solar panels involving six layers of different materials each tuned to a different wavelength of light achieved an efficiency of 47 percent in the US National Renewable Energy Laboratory.[9] Experiments like these are encouraging, but there is always the trade-off between complexity and cost. More complex materials usually cost more while cheaper substances may not be as efficient. It is an ongoing process to come up with the optimum solution.

One criticism of silicon cells is the very high temperatures needed to turn the mineral into a single pure crystal, which, at least in current Chinese manufacturing, means using fossil fuels, usually coal. So solar energy is not entirely carbon free.

Considering the carbon released during manufacture, around 50 grams of CO_2 per kWh is produced by solar panel systems during the first year of operating. That is roughly twenty times less than the carbon footprint of coal-powered electricity. After three years of operation, the solar panels will make up for that debt and become carbon neutral, remaining that way over a twenty-year lifespan. As more clean alternative sources of electricity come online, that carbon footprint will be reduced much further until, potentially, solar panels are being made by energy from solar panels.

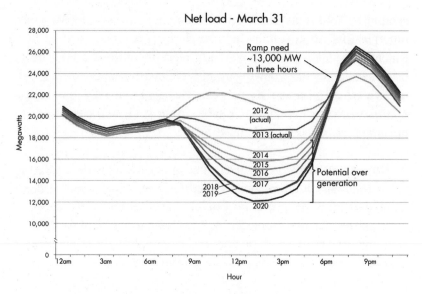

Graph of energy demand during a day, the so-called duck curve. The top line is power output from solar energy.

Licensed with permission from the California ISO

A big elephant—or actually the big duck—in the room is the fact that the sun doesn't always shine, and when it does, it is not necessarily when we need extra energy. The phenomenon is known as the duck curve.

If you think about household energy use, it is low overnight, rises in the morning as people turn on their appliances to make breakfast and turn up the heat in winter, goes down during the day as they leave home to go to work or school, and rises again in the evening when they come home for dinner and turn on the evening's entertainment. Plot that rise and fall of energy on a line graph and it looks like a duck.

On the other hand, if you think about solar energy availability, it is weak in the morning when the sun is low on the horizon, peaks at midday, and falls off in the early evening—the opposite of household demand.

Furthermore, when the peak power of solar energy during the middle of the day is added to the grid, the demands on conventional power generators such as coal-fired plants, hydroelectric dams, or nuclear plants is reduced, and the value of that conventional electricity goes down. These big generators cannot be turned up and down easily like a thermostat; they are most efficient when running flat out. The exception is generating stations that run on natural gas, which can be dialled up and down quickly and, therefore, often accompany wind and solar farms to compensate for the slack times. Many argue that the benefits of clean alternatives are negated by the gas plants. Why not just have gas alone? But gas-fired plants are still burning fossil fuels, and while they are cleaner than coal or oil, they still produce carbon dioxide, a greenhouse gas that warms the climate.

No matter how efficient the solar cells are, they are operating below their peak performance most of the time. They, of course, don't work at all at night, and when the sun is up, the amount of sunlight falling on the Earth varies throughout the day as well as with the seasons and with latitude. Their operation is also affected by clouds and dust in the atmosphere, pollution, and even dirt accumulating on the surface of the panels themselves. This intermittent power doesn't fit with our constant energy demands where we expect the lights to come on when we flip a switch no matter what time of day it is.

Electricity is an energy carrier, a means of moving energy from one place to another. It is not a source like oil that lies around waiting to be used. Electricity is so volatile that we have to use it as soon as we make it. The lights in your room right now are powered by electricity that was probably generated just a few minutes ago by a generator in your region. It is hard to hold on to and store electricity the way we can hold oil in huge tanks for later use.

Thankfully, the laws of nature have provided a way out of this problem. The law of conservation of energy states that energy cannot be created or destroyed, only converted from one form into another. So the solution to storing electricity is to turn it into another form, whether it be the chemical energy in batteries, the potential energy of falling water, the mechanical energy of machines, or some other innovative energy conversion. This is the subject of a later chapter.

The holy grail for solar energy is highly efficient cheap materials combined with cheap energy storage. If that can be found, solar energy could come to dominate in the near future. Finding large tracts of land for solar farms is not as difficult as it seems, especially when compared to other ways we convert land to our purposes.

Photo rendering of Project Nexus, a test of solar panels over canals in California, announced February 2022.

Image courtesy of Solar AquaGrid LLC (www.solaraquagrid.com)

According to the United Nations, 12 million hectares of forest are destroyed each year.[10] If we built solar farms at that same rate, the world would have enough power for everyone in three years. Fortunately, there are other approaches to finding places for solar panels that do not involve cutting down more forest. For the United States, the amount of land needed to become self-sufficient on solar is roughly the area of the interstate highway system, which was a national megaproject built in thirty years.

This is a matter of engineering and political will. Tesla, Inc. already manufactures a complete package of solar roofing tiles and battery packs for the home and has provided huge mega-pack batteries to back up some of the world's largest solar farms in Australia and elsewhere.[11] For sure, there will be large-scale solar projects in desert regions such as the southwestern United States, the Middle East, India, and Australia, where sunlight is plentiful more than three hundred days of the year. But while large desert projects can produce a lot of power, people don't live much on deserts, so the challenge is getting that power to major centres that can be hundreds of kilometres away. The farther power is transmitted, the more loss to resistance in the lines. In response, some regions are finding clever ways to build large-scale solar farms closer to cities without taking up more land. They build them over water.

India is investing heavily in solar power under the slogan, "One Sun, One Power, One Grid." It has provided incentives to make solar panels common on rooftops, including airports and railway stations, and is building some of the largest solar farms on the planet. But the country depends heavily on agriculture to feed a population of more than a billion people, so there is reluctance to give up that valuable land to farming sunlight. Still, the Indian government is investing heavily in alternatives to get itself off coal. It is building large solar farms such as the 2,050-megawatt Pavagada Solar Park in the southern region, which has been drought stricken. The power companies provide income to the farmers by paying them to lease their land for twenty-five years. The enormous park covers fifteen square kilometres, or 13,000 acres, and provides power for roughly a million homes.[12]

Beyond the megaprojects, India is taking another very innovative approach to incorporate solar into existing structures without occupying more land. About half of India's agricultural land is irrigated through a vast network of canals that have been built over the centuries. These canals run for tens of thousands of kilometres across the country.

In 2014, a 750-metre pilot project launched in the state of Gujarat covered canals with solar panels. The solar power ran irrigation pumps and, as an added benefit, retained moisture in the canals by cutting down on evaporation, making more water available for agriculture and drinking. A third benefit came from the efficiency of the solar panels themselves. Ironically, solar panels tend to lose efficiency when the sunlight makes them too hot. But by placing the panels over canals, the cooling effect of the water kept the panels at a more efficient operating temperature and increased their lifespan. Finally, the shading cut down on harmful algal blooms caused by sunlight that can make water toxic. It was a win-win-win situation all around. Since then, 40 kilometres of solar panel–topped canals that branch off the Narmada River have been commissioned with a potential electrical output of 100 megawatts, with more likely to follow.[13]

California, which has 6,350 kilometres of irrigation canals and has been subject to numerous droughts, is considering the same option. Studies have shown that the savings in water evaporation alone can be 40 to 90 percent.

These examples of incorporating solar power into a community without sacrificing land are one way solar energy will advance in the future. It will become more locally produced and often invisible, hiding in windows, the cladding on building exteriors, or your clothing, gathering sunlight wherever possible to keep our devices charged and the lights on in our energy-hungry world.

It is all there for the taking.

So what does this mean for the average consumer who is considering investing in solar power?

It depends on where you live and where your current power comes from. Do you live in a region that gets a lot of rain or in a dry area with many days of sunlight? If your power comes from coal-fired generating stations, is it worth considering solar for environmental reasons?

Solar energy systems require either batteries for backup or a connection to the grid so you can get power at night. But batteries are expensive. An alternative is to contact your utility company to see if they will accept two-way meters, so that you receive a credit when your house is producing excess power that feeds into the grid and can tap into it at night. This is essentially using the grid as a huge battery, although a small backup battery pack will keep you running during power outages.

Look at your monthly utility bill to see how much your home consumes to determine how much solar energy you need. Then talk to an energy consultant to choose the best design.

The initial investment in solar may seem high, but many systems pay for themselves in savings in four to six years, usually with guarantees that the system will run for twenty to twenty-five years. Take advantage of government subsidies, tax rebates, and other incentives to reduce costs.

Installation can be done by turnkey companies such as Tesla, which will provide all the equipment and install it for you. But that comes at a premium. You can also purchase the equipment from a wholesaler and have local contractors, such as roofing companies, install it at less cost. It will take a lot of research ahead of time to ensure that all your components will actually work together as a single system. Make sure you work with an expert before getting yourself into an expensive bind and having to start over.

The sun has roughly another four billion years of life left in it. That's a lot of free energy streaming our way. The technology to capture sunlight and turn it into useful products and services is evolving rapidly and becoming more versatile. How we deploy those systems is a personal, political, and social decision as we move towards a decarbonized world.

So what will that future energy world look like?

I'm a bit prejudiced, but I think solar will be the biggest prime in the long term, followed by wind energy, then geothermal, which has big potential that is still to be explored, especially for low-temperature heating purposes. Then there will be smaller contributions from other renewables, but from a technological point, based on where we are already, solar and wind are going to be pretty hard to beat.

Solar's role and contribution also depend on further developments of storage and conversion, such as large-scale electrolysis, cheaper batteries in different forms, and heat pumps, but as an energy source, or energy generator, it is here and growing rapidly.

"Solar is the new king," and that comes from the International Energy Agency, which is a kind of historic statement, I think. But that just confirms what the people in the sector already know.
Wim Sinke

The future of solar power looks very bright. It has the greatest potential of all clean alternatives to decarbonize our energy system. According to a study by the Energy Watch Group in Germany and LUT University in Finland, global demand for electricity is expected to rise by 1.8 percent every year due to modernized lifestyle incorporating more electronic devices, heat, transportation, and desalinization. The group projects that Solar could supply more than 70 percent of that electrical supply by 2050.[14] It is no longer a matter of technical or economic feasibility. Green technology is one of the fastest growing economic sectors with the potential to provide jobs as it spreads around the globe.

The technology is mature, with a long history of mass manufacture, making it the cheapest and most versatile clean energy source of all. It can be scaled up from rooftop installations that power your home to giant solar farms covering many square kilometres supporting a city. With new advances in thin film solar, electricity will be coming from windows, walls, sculptures—places you won't even notice—as it becomes incorporated into architecture.

Our current fossil fuel–based energy that powers our homes and workplaces has travelled distances to reach us. Perhaps it was produced in a large generating station outside the city limits, relying on pipelines laid across thousands of kilometres or oil-rich countries halfway around the world.

Solar energy does not have to be imported from somewhere else; it can come literally from your own backyard. For countries that have no fossil fuel resources of their own that means energy independence. It also reduces international conflict over limited energy supplies.

To break away from centralized power generation will involve more than simply installing more solar arrays. It means levelling the playing field by eliminating subsidies to fossil fuels and giving tax incentives and investment opportunities to clean energy. It also means converting our energy consumption towards more electrification, as we are already seeing in the automotive industry. But it needs to be carried further to other energy consumption, such as switching gas or oil furnaces to electric.

There is a strange irony in that the sun has become somewhat of an enemy as the climate warms. It ignites wildfires, dries up lakes, causes droughts, bakes our cities, and burns our skin. Yet that same sunlight can help prevent further warming of the climate by decarbonizing the atmosphere. The sun can either kill us or provide us with the greatest source of limitless free energy. All we have to do is find innovative ways to put it to work.

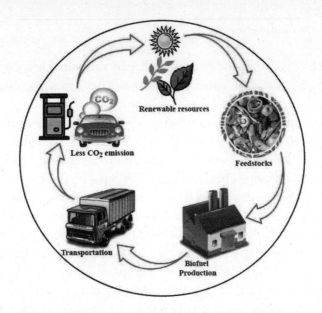

Biofuels have many uses and the feedstocks come from various different sources.

Image courtesy of "Recent advances and viability in biofuel production," *Energy Conversion and Management*: X, Volume 10 (2021) 100070

Chapter Two

Biofuels

Our first fuel, now energy from waste, forests, or crops. Are they sustainable?

Picture the scene: a Neanderthal family around a fire in their cave, and one of them asks, "What's on tonight?"

Another replies, "Well, at seven o'clock it's Birch, seven-thirty we switch to Pine, and at eight we have a special feature with Oak."

Our human ancestors and ancient relatives were the first users of biofuels. Burning wood has been central to human development and survival for the last few hundred thousand years or longer. Wood fires enabled people to survive colder climates, cook food, and, along with coal and oil, ultimately drive the Industrial Revolution. Now we can enjoy the romance of an open fire by throwing on a log and sitting back with a fermented beverage, another technology also thousands of years old, which has now become a biofuel to drive vehicles.

Biofuels are stored solar energy. That flickering fire is the release of sunlight that was captured by leaves through photosynthesis, stored in the living tissue, and then, when burned, turned back into energy in the form of heat and light. The burning also releases carbon dioxide, which was absorbed by the leaves and stems when the plant was alive, along with pure carbon that rises up as smoke and dissipates in the air. For most of our history, nature, the perfect recycler, would absorb that carbon dioxide from fire into the leaves of new plants growing in place of those that were burned, while the pure carbon returned to the soil. As long as that carbon cycle from plants to the atmosphere and back again was balanced, the CO_2 levels in the atmosphere remained fairly constant. Occasional periods of volcanic eruptions or meteor impacts would disrupt the balance for a time, but nature would always eventually restore order.

Today, we are going beyond the campfire by converting organic matter into other forms of energy to generate electricity or run vehicles, and then we try to recapture carbon emissions from that combustion with new plants to complete the cycle. Ideally the entire process is carbon neutral. But in reality, carbon neutrality is seldom achieved because energy must be used to plant and harvest the crop, then more energy to transport it, and more yet to convert it into a useful energy source. Bioenergy still involves carbon emissions. However, those emissions are lower than what would come from burning fossil fuels.

According to the US Department of Energy:

> Each gallon of corn ethanol today delivers as much as 67 percent more energy than is used to produce it. In the future, most ethanol will come from cellulosic ethanol (wood from forestry waste products or wild grasses), which delivers up to ten times more energy than is required for its production.[1]

Extracting energy from living things involves three distinct terms, all with their own issues:

Biomass is the feedstock, anything made from organic material, either alive or recently alive, whether it is wood, corn, canola, sugar cane, or waste materials such as garbage or even sewage. Where the feedstock comes from is an important consideration. Is it, for example, coming from sustainable farming practices or from destructive clear-cutting of rainforests to make way for palm oil plantations, such as in Borneo?

Bioenergy is taking that organic material and converting it into some form of useful energy, such as heat to generate electricity or liquid fuels to use in transportation.

Biofuels are solids, liquids, or gases that can be used to replace fossil fuels. For example, wood pellets can replace coal in generating stations, ethanol added to gasoline reduces the amount of gas being used. Then there are the so-called drop-in biofuels that can be run directly in trucks or aircraft with no modifications to the engines. Natural oils such as canola or used cooking oils are processed into biodiesel or jet fuel that is indistinguishable from the fossil versions.

While biomass, bioenergy, and biofuels all involve natural products, there are questions surrounding every aspect. Where is the original feedstock coming from, and is it sustainable? Does it come from managed forests of fast-growing trees, sustainable sugar cane plantations, or virgin forest? Are fuels made from farm waste such as manure or sourced from waste treatment plants, or are they derived from food crops such as corn, which could be used to feed a hungry world? And finally, what overall impact is biomass consumption having on the natural environment? For example, natural forests are very diverse, containing many different species of plants and creatures that live among them. Plantations tend to be monocultures with much less diversity. According to environmentalists, we need to "rewild" the Earth, dedicate large areas where natural forests can grow back.

In some parts of the world, biofuels are a sustainable industry. Other places need to address these issues more seriously.

Does that mean we can just grow our way into a clean future?

Obviously, biofuels cannot totally replace fossil fuels. Humanity's voracious appetite for energy would burn up everything on the planet. But they can play a role as a transitional fuel and fit into niche areas where sustainable agriculture can support it.

The modern biofuel industry began during the oil embargo of the early 1970s when gas prices shot through the roof. The two biggest producers of biofuels were the United States and Brazil. According to Dr. Jack Sadler, University of British Columbia and International Energy Agency Bioenergy Task 39 BC-SMART program,

> The US used corn to make ethanol and Brazil went sugar cane to ethanol. That is a relatively straightforward process. We have been doing it for millennia in terms of making beer and wine.
>
> The terminology that might come up is first generation, second generation, third generation. And the Brazilians get a bit upset because they were first generation, but Brazil grows sugar cane quite well. In terms of carbon intensity, it's not that bad, so the ethanol that they produce has been getting steadily better. The electricity they use is mostly hydroelectric.
>
> The US is a bit more of a biorefinery approach because the sugar comes from the starch. You also have the protein. If you look at the dry mill and the wet mill, companies like ADM have done a good job of not only lowering the cost of ethanol but also offering multiple products. If you look at the various products that come out when you've got corn as the feedstock, they've been the big guys.
>
> So, per hectare of land, ADM is making sugars pretty well.[2]

ADM is a multinational company that began as a linseed oil producer in 1902 and has since diversified into producing a variety of food products—from health and wellness, flavour ingredients, and beverages to animal and pet feed—along with biofuels, as well as managing farming operations while not owning any farms itself. Their operation shows how biofuels can easily fit into existing agricultural processing.

Biomass Processes

Modern biomass production involves five different processes. The first is direct burning by just cutting down a tree and throwing a log on a fire. This is still important in many parts of the world today, especially in developing countries where people are having to travel farther and farther to find wood to use as their only cooking fuel. There is also a wood pellet industry where wood is pressed in cylindrical mini-logs about the size of your thumb to be used in home wood stoves and large-scale generating stations. In some northerly regions, peat can be dug out of the ground, dried, and burned directly.

The second method uses a very old technology that goes back to the ancient Egyptians and Babylonians who brewed wine and beer more than five thousand years ago. Rather than turn grapes into wine, we now start with crops containing sugar, such as corn, grains, or sugar cane. Add yeast, let it ferment for a while, and draw off the ethanol, which is the same kind of alcohol we drink. Mix it at 10 percent with gasoline and you get a blend of gas known as gasohol, which can be run in conventional engines. The addition of the biofuel means there is less gasoline being used, which reduces overall carbon emissions.

The third process uses extracted vegetable oil from plants such as soybean or animal fats that are heated and chemically processed to produce a liquid very similar to diesel fuel.

Fourth on the list is digestion, where lower-quality organic material such as farm or human waste, corn husks and cobs, or grasses are put into a large vat, oxygen is kept out, and anaerobic bacteria go at it as food. The methane gas released in the process can be burned as a replacement for natural gas.

Lastly, fast-growing microscopic algae, the kind that forms green scum on ponds, can be grown in open pools or indoor glass tubes and harvested every day. Fats, or lipids, can be processed out of them and turned into liquid fuels.

Biomass comes with two big selling points. First is the renewability, though it can only be achieved if the same, or a greater, number of plants are grown than are harvested for fuel. Second is that if turned into liquid and gas fuels, biomass is the closest replacement for fossil fuels that can be run in conventional furnaces and gas, diesel, or aircraft engines without modification. Energy derived from biomass can work well in regions that do not have their own fossil fuel reserves but do have lots of agricultural land by reducing dependence on costly imported fuel.

Ironically, the oil industry supports biofuels even though the products are intended to replace oil. Our consumption of gas and diesel is so high, about 100 million barrels per day worldwide, that biofuels alone cannot keep up, which is the main reason they are blended with fossil fuels. It enables the oil industry to continue production. The auto industry is also happy because they can continue business as usual with no major changes to the technology of combustion engines. It seems like a win-win for all.

Biofuel Processes

While biofuels will reduce the amount of fossil fuels being burned, the question is whether the biosphere can keep up as more combustion vehicles are added on the road. Can plants grow back as fast as we are using them?

How long does it take a tree to grow? You're looking at seventy-five to one hundred years, so what is the payback period?

And in the short term, they're saying the burning of pellets is not as efficient as burning coal. The problem with any of the fossil fuels, even natural gas, which is brought out as the good fossil fuel, is that, unfortunately, CO_2 is still going up there.

If you tend to think of it generationally, it will take some time. However, in somewhere like Brazil, they're good at growing trees pretty fast, so they can sequester that carbon back out of the atmosphere quite quickly. A young tree sequesters carbon faster than an older tree, kind of like a human being. I'm still sequestering too much carbon by drinking too much beer.

It's quite contentious, it's not black and white, and it's all going to be site-specific in terms of how well the forest is being managed.

Dr. Jack Sadler

The following is a breakdown of the different biofuel processes.

WOOD PELLETS

The simplest use of biofuels is to simply burn it the way coal is burned in generating stations. In fact, wood, when cut down into pellets, can be mixed with coal or replace it altogether in conventional generating stations. This reduces or eliminates harmful sulphur dioxide and nitrous dioxide emissions produced when burning coal alone. Canada is a large producer of pellets because of the huge areas of forest and the large forestry industry that provides materials that would otherwise go to landfills. Most pellets are made from the by-products of the lumber and paper industries, in other words sawdust. They are also made from the low-quality trees of no economic value and left over from forestry operations and, sadly in more recent times, from millions of trees killed by the mountain pine beetle infestation in North America.

Wood pellets are made by chipping wood then adding heat to dry it. When moisture content is below 12 percent, the chips are ground into a powder that is then compressed into small pellets that are easy to transport and store. These pellets have more than three times the energy density of regular wood and, when burned in high-efficiency pellet stoves and boilers, can have a combustion efficiency of up to 85 percent.[3] According to the Wood Pellet Association of Canada, wood pellets emit 90 percent fewer emissions than coal and 50 percent fewer than natural gas.[4]

Wood pellets for fuel.   Photo courtesy of Wood Pellet Association

One of the largest wood pellet generating stations is operated by Drax in Selby, North Yorkshire, in the UK. It hosts four wood pellet–fired units totalling 2,595 megawatts, producing 14.1 terawatt hours in 2020, making it Europe's largest renewable energy generation site.[5] The company is planning to incorporate carbon capture and storage of the emissions from the power plant so it will be more than carbon neutral. It could become carbon negative because the carbon absorbed by new trees growing would be greater than that emitted by the generating station.

Obviously, countries with large tracts of forest, such as Canada, Russia, and the United States, are leaders in wood pellet supply. But the carbon emitted by burning wood products will not always be absorbed immediately by new plants because of the time it takes for them to grow. So, in the short term, emissions can be greater than absorption.

> Bioenergy is not for everyone. It's not for Japan or Korea. But if you look at wave and tidal, it's for the Japans and the Koreas. So who's focusing on bioenergy? It's the Scandinavians, the Canadians. The focus on these different areas will depend on the assets of those countries.
> *Dr. Jack Sadler*

ETHANOL

You have probably noticed labels on gas pumps that say, "May contain up to 10 percent ethanol."

These are blended biofuels where gasoline is mixed with ethanol produced through fermentation. All that is required is plants containing sugar, yeast, and a little time. But rather than the grapes, barley, or hops used to make drinking ethanol in the form of wine and beer, the stock plants for the world's largest producers of fuel ethanol are corn in the United States and sugar cane in Brazil. In fact, farmers in the US have found the value of their corn increase because of government regulations to incorporate ethanol into gasoline. Today, about 40 percent of the corn and maize produced in the country is used for fuel.[6]

The technology for converting sugars to alcohol has been around for centuries, so it is certainly a mature method. But in terms of carbon emissions, the amount of fossil fuel needed for machinery used during planting, harvesting, and processing of the fuel must be calculated in to determine how much carbon has actually been prevented from entering the atmosphere.

Another approach to ethanol production is to use the waste products from agriculture rather than the main crop itself. However, this must be balanced against the ancient agricultural practice of leaving material behind to replenish the soil.

> The US Department of Energy looked at corn stover, so their feedstock was agricultural residue. A lot of work went into how much you need to leave on the soil to keep it healthy, and they are on the side that there is more than enough. You could probably take about half of that residue off the land without it decreasing the amount of phosphates or anything else.
> *Dr. Jack Sadler*

A third alternative feedstock for ethanol fuels is switchgrass, which is fast growing, not used for food, and native to North America.

BIODIESEL

Go to the cooking oil section of a grocery store and you will see a line of vegetable oils derived from myriad different plants—olives, linseed, peanuts, the list is long. Any of these plants and seeds could be turned into biofuel, although the rarer ones would not be able to keep up with the demand. The best candidates are those that can be grown on a large scale, such as canola. These oils can be either blended with regular diesel fuel or burned directly, as folks who run their diesel pickup trucks on used french-fry oil from restaurants know. (The trucks even smell nice when they run.)

In North America, soybeans and canola are popular feedstocks. About half of the soybean crop is crushed into oil and only about one-fifth of that goes into biodiesel, so about 10 percent of the total crop is used for biodiesel.[7]

A more purified form is known as HVO/HEFA, which stands for Hydrotreated Vegetable Oils/Hydroprocessed Esters and Fatty Acids.

The biodiesel produced is almost indistinguishable from regular diesel but lacks the aromatics and sulphur content that are responsible for pollution.

The big player in this industry is Neste Oil, which has facilities in Finland, Singapore, and Rotterdam producing biodiesel from a number of different vegetable oils, including palm oil.[8]

However, we burn much more diesel than we plant crops or eat french fries, so all of these vegetable oils only make up about 4 percent of diesel fuel.[9]

Biofuels derived from natural oils can play a big role in reducing emission for long-distance transport, such as trucking, rail, marine, and air, where electrification is not yet practical.

Make Biodiesel at Home

Start with pure soybean oil, which you can mix with other oil, such as french fry, if you choose. Use a heat-resistant container that can be drained from the bottom.

- Heat 800 ml of oil to 54°C.

- Add 6 mg of lye, or potassium hydroxide. This adjusts the ph level.

- Add 175 ml of methanol (not the ethanol we drink). This produces the diesel fuel as well as glycerine, which still contains methanol and can be recovered.

- Let the glycerine settle out.

- Drain the glycerine from the bottom. What remains is diesel fuel.

BIOGAS

This is where biofuels become innovative because it deals with materials we are normally trying to get rid of: organic waste. And it is done by letting microbes do the dirty work for us. Biogas is the by-product of bacteria that break down material into methane and carbon dioxide. In other words, it is the same process that happens in nature when something dies and rots or, to put it bluntly, when you fart. It's the bacteria in your gut breaking down your food that produce the embarrassing gas as a by-product.

Wastewater treatment plants are an extension of the human digestive tract. The waste that we deliver from our toilets is put into large digesters filled with bacterial colonies that are quite happy to treat it as food. The digesters are anaerobic, that is, sealed off so there is no oxygen in the system, and biogas, mostly methane, bubbles to the top of the chamber where it can be drawn off and burned as fuel. Many waste treatment plants are actually run on their own biogas. That methane can also be combined with carbon dioxide to make syngas, which burns like gasoline.

The same process works for other organic products such as food waste or waste from agriculture, such as cane stalks, corn husks, or fast-growing plants such as switchgrass.

ALGAE FUEL

If you have been to a pond or small lake in late summer, you may have seen a green slime floating on the water or attached to plants along the bottom. That pond scum is algae, one of the oldest forms

Algae farm with long channels filled with water containing green algae.

Image courtesy of iwi

of life on the planet. They are actually green plants that use sunlight to grow, but each is a single microscopic cell. Those cells reproduce incredibly quickly, doubling their numbers within hours, forming vast colonies that we see as an algal bloom. When cultivated on a large scale, they can grow twenty to one hundred times faster than conventional crops and thus hold great potential as a feedstock for biofuels.

The key to these biofuels are lipids, which are fatty molecules that store energy and supply structural support for living things. Algae are up to 40 percent lipids by weight so they are the super lipid suppliers.[10] Lipids can be processed into diesel, synthetic petroleum, butenol, or industrial chemicals.

Commercial algae farms use large outdoor ponds in the shape of long interconnected troughs with paddles continually stirring the liquid along like artificial river rides in amusement parks. This keeps all the algae exposed to sunlight to promote growth. When the density of algae in the water reaches a critical level, it is harvested, dried, and then mixed with solvents to remove the lipids, which are then mixed with hydrogen to make the fuel. It's a complicated process that takes energy, making this fuel about three times more expensive than diesel.[11] Algae farms were very popular in the early 2000s, but companies attempting to produce fuel on a commercial scale did not meet their targets and had to move to other algae products such as food supplements, cosmetics, and animal feed. However, in places such as the Pacific Northwest National Laboratory in Seattle, research continues to lower costs by developing new strains of algae as well as extracting other products from the algae such as phosphorus, which is used in fertilizers.

BIO-JET FUEL
Most biofuel production has been focused on ground transportation, but now the aviation industry is chiming in to reduce its carbon emissions as well. According to the International Energy Agency, aviation will account for 3.5 percent of global CO_2 emissions by 2030, despite ongoing improvements in engine efficiency.[12] The aviation industry has committed to reducing carbon emissions by 50 percent from their 2005 level by 2050. Jet fuel, which is mostly

kerosene, has been the only fuel available for airline travel, and any alternative must exactly meet the same performance requirements. Several approaches, involving almost all the above methods from soybeans to algae, are employed to produce sustainable aviation fuel (SAF), which are biofuels blended up to 50 percent with jet fuel to meet that goal. These fuels must be thoroughly tested in laboratories and flight tests to ensure that they perform exactly like pure jet fuel with no changes to the jet engines. So far, that performance is meeting expectations; however,

Aviation will account for 3.5 percent of global CO_2 emissions by 2030. The industry is incorporating biofuels to reduce carbon emissions.

© Jennifer Hartley

the biofuels are more expensive than jet fuel, and fuel is the highest cost to airline operations. That cost could come down if fuel is made from cheaper waste materials.

Many airlines, including United, Virgin Atlantic, KLM, British Airways, and Lufthansa, have committed to using biofuels, especially on long-haul routes. At the same time many international airports have begun offering biofuel, which can be added to fuel storage tanks, again without modification. Oslo Airport led the way in adopting SAFs in 2016, with United Airlines at Los Angeles International Airport close behind.

Still, aviation biofuel production of about 15 million litres in 2018 accounted for less than 0.1 percent of total aviation fuel consumption. This means that significantly faster biofuel production is needed to deliver the levels of SAF required by the aviation industry. Blending lower-carbon SAF with fossil jet fuel will be essential to lowering carbon emissions from airliners. This is reflected in the IEA's Sustainable Development Scenario (SDS), which anticipates biofuels reaching around 10 percent of aviation fuel demand by 2030, and close to 20 percent by 2040.[13] This may be the best use for biofuels in the future because there are few alternative fuels for jet engines.

Seeking a Sustainable Balance

The rise in demand for biofuels by both the aviation and ground transportation industries presents a conundrum. On one hand, biofuels reduce the amount of fossil fuels consumed and therefore the amount of carbon added to the atmosphere. On the other hand, their production takes up agricultural land that could be used for food production, and in a world where billions of people are hungry, burning food in our vehicles seems unjust. In addition, as demand for biofuel rises, more natural habitats will be covered with agricultural land.

The cost of biofuels must compete with the cost of fossil fuels to be economically viable. This is difficult in light of increased oil drilling and fracking technology that has kept fossil fuel prices relatively low. And when those prices do rise, so too does the price of, for example, corn, because fossil fuels are needed for planting and harvesting equipment, which means people end up paying more for food.

So where do we draw the line in this food/fuel debate?

I look at Malaysia and Indonesia. What Malaysia is doing is palm oil. And there is some good palm oil because it has been growing there for generations. It's about raising the standard of living, but you don't want to raise it to where everyone is eating steaks all the time. There is a growing realization that that is not the way you want to go. The balance between food and fuel is about what kind of food we want to eat. It is multilayered and complex, and I don't think it will ever go away.

If you are filling up your SUV with renewable diesel, someone's going to say, that's enough feedstock there to feed a family for a year. That type of question will never go away.
Dr. Jack Sadler

In countries such as Borneo, vast areas of rainforest are being cleared to make way for palm oil plantations and, at the same time, are destroying the habitat for orangutans, one of our closest primate relatives. Then there is the issue of water use. Agricultural biofuel production takes water away from food production. By comparison, ethanol biofuel has a water footprint on the land hundreds of times larger than the extraction of fossil fuels,[14] including water-intensive oil sands operations. Given that droughts are part of the climate change scenario, water is becoming a scarce resource.

In 2020, the United States burned 2.956 billion barrels of gasoline.[15] That's 338 million gallons per day or more than a billion litres every day just in the US. Biofuels can never completely replace that volume of fuel consumption, which continues to rise.

Currently, in the US, out of the total corn crop, 23 percent is dedicated to ethanol, 37 percent to animal feed, 15 percent to food, 14 percent to exports, and the remaining 11 percent goes to other corn products.[16] (These figures vary from year to year.) The largest production goes to animal feed so we can eat meat. Therefore, corn will likely not be able to provide more than a 10 percent blend for ethanol in gasoline before it disrupts the food supply and adversely affects food prices. On top of that, biofuel production costs depend on crop yields and market prices, which change from year to year.

From the farmers' point of view, growing corn requires an investment in fertilizers and in equipment for ploughing and harvesting. According to a study out of the University of Illinois, the net social and economic worth of corn for food in the US is $1,492 per hectare verses a $10 per hectare loss for biofuel corn production.[17] That is why biofuel is subsidized by the government. However, as the demand for ethanol rises, the value of corn is increasing for farmers. The report also says that the depletion of nutrients from the soil that has to be replaced by fertilizers negates any environmental gains. To meet the demand for biofuel, more land and more food crops will be taken over for energy use.

Sustainable biomass energy production depends on the type of agriculture or forestry. A set-up engaging local farmers could benefit the local community by increasing agricultural production and offering new economic opportunities. But the value of biofuel has to be worth it for the farmers and foresters.

What you need to look at is, instead of drilling a hole and getting natural gas, oil, or coal, biofuels are much more tied into the foresters and the farmers. We've got to ensure the sustainability of that feedstock, and there is already a supply chain for those materials.

Why do we grow crops and grow trees? We use it for pulp and paper, for solid wood, so really, that kind of energy source has not been a dominant market in the past; however, it is growing.

The problem is, it tends to be low value. If you look at corn, that corn is probably more valuable to feed the cows than it is to make ethanol. If you look at the forestry sector, if you make a two-by-four or packaging material for Amazon, then make energy out of it, there is lower value. That is the pressure on the current sector, because they are going after the higher value products. It's good that they use the residue and the waste.

It has to be market driven. That's why policies like the carbon tax and the carbon fuel standard that we have in BC and they have in California are good policies to move us that way.
Dr. Jack Sadler

Finally, there is one more benefit to our natural resources that goes beyond providing products we can burn: the value of the biosphere itself. A mature forest is a balanced ecosystem that is home to myriad plants, animals, insects, and microbes that thrive together in an intricate interconnected web. As the famous environmentalist John Muir, who established the national parks system in the US, said, "When one tugs at a single thing in nature, he finds it attached to the rest of the world."[18]

As we enter these natural habitats with saws, bulldozers, and trucks, we are disturbing a natural balance. Cutting down one tree disturbs the lives of millions of other life forms, from mosses on the bark, to insects and birds in the canopy, to bacteria in the soil and fungi on tree roots. Those roots not only hold onto soil and prevent erosion but also, we've come to learn, form an underground biological communication network between trees. A single tree is an entire ecosystem within the larger ecosystem of the forest.

A diversified forest is a healthy forest, where disease or fire may affect some species but not others. Microbes and insects break down leaf litter and fallen logs, returning nutrients to the soil, while larger animals use the dense vegetation as cover in their endless game of predator versus prey.

A forest is a sanctuary for the human spirit where people can escape the hectic lifestyle of cities and find a tranquility that is good for mental and physical health. That is a value that cannot be measured in dollars.

As more pressure is exerted on forests to provide products for our industrial world, will we be able to leave some of them alone?

> Yes. I think of growing up, and I had my Sunday clothes and my weekday clothes. I look at consumers now and clothes are disposable, furniture is disposable. But I think there is increasing recognition that's not sustainable. And working at a university, I've got great hope because I look at the young people I'm lucky enough to work with. I think this era of consumerism that I grew up with is being balanced with quality of life. The kids are saying that it's not about going out and buying a big car. It's about quality of life, so I'm optimistic.
> *Dr. Jack Sadler*

In the end, biomass is not a complete solution to our energy and climate problems. Our fuel consumption is simply too great for plants to satisfy our thirst. (As the saying goes in real estate, "They're not making any more land.") In reality, biofuels are helping the old technology we are trying to get away from, such as the internal combustion engine, stay around longer. They will remain viable as long as combustion engines running on gasoline and diesel continue to dominate the roadways and oil prices continue to rise. The role of biofuels is as a transitional technology that will buy us time while our transportation sector evolves into another form that does not require fossil fuels at all.

Chapter Three

Wind Power

How the three-bladed design came to be the best to meet our energy needs.

A dark squall line hovers over the city of Miami. I'm sailing along the Florida coast for the first time and have seriously underestimated just how strong afternoon storms become and how quickly they sweep through the area. All sails are up, which is a bad situation in a storm. A wise captain shortens sail before a squall arrives so the boat will not be overpowered. In a matter of minutes, the beautiful, calm, blue-sky day turns into a maelstrom of wind and rain, while a powerful invisible hand tries to push my sails into the sea.

Sailors have a love-hate relationship with the wind. Cruising along in a moderate breeze with sails unfurled, forming beautiful curves, the hull silently and gracefully gliding through gentle swells like a dolphin:

it is the perfect balance between wind and waves. But when those waves rise above the deck, streaked with white foam whipped up by the wind, it becomes a battle for survival as nature tries to swallow the ship into the sea without a trace. In sailing, there is such a thing as too much wind.

This was my first time sailing in tropical waters and I had seriously underestimated how quickly storms move in these regions. A dark squall line appeared on the horizon. My inexperience made me completely unprepared to face the force of nature.

Skies swiftly darken overhead. I watch dumbfounded as the needle on my wind speed indicator climbs in a matter of seconds like the speedometer in a fast car . . . 15, 20, 25, 30 knots and still moving. Sail fabric strains against the rigging as more than 15 tons of boat is forced over onto its side. Railings along the outer edge slide under the water as sea foam washes over the deck. The rain arrives just as suddenly, as though someone has thrown a bucket of water in my face. Whipped up by the wind, the ocean churns, the surface turns white. Large waves appear out of nowhere. I release the sails to dump the air and struggle for control. They flap like giant flags dragging in the water. The boat rights herself slightly, but we are still being blown sideways like a leaf across the surface. I look up to see mountains of green water that lift the entire boat as they roll past then drop us into watery valleys, a roller-coaster ride between white-capped peaks. I wonder whether my first sail in the tropics will be my last.

People have been hanging large pieces of fabric on sticks to capture the wind for millennia. It enabled ancient mariners to cross seas, and for those on land to pump water out of the ground. In fact, wind power was a more common technology in the past than it is today. Harnessing the wind for transportation goes back to 5000 BCE, when sailing craft ran people and goods up and down the Nile. Around 200 BCE, wind-powered pumps were used in China and by the eleventh century, wind-powered mills were widely used for food production throughout the Middle East. The technology made its way north to Europe where the Dutch developed their iconic four-bladed structures, several storeys tall, to drain the land and keep water behind dikes. You could say that wind power built Holland, for without thousands of those wind-powered pumps, the country

would be underwater. Of course, they were also the favourite target of Don Quixote.

Wind is converted solar energy. The Earth's atmosphere is a giant heat engine, where nature very effectively takes light from the sun and warms the ground, which heats the air and turns it into the kinetic energy of the wind. If you think of the Earth as a ball floating in space, the sun is off to one side, so the planet is heated unevenly. The middle of the ball along the equator gets more direct sunlight than the top and bottom. This creates a dance of dominance between warm tropical air and cold polar air that continuously attempt to change places, the heat moving towards the cold, the cold rushing in to take its place. Coupled with the turning of the Earth, our atmosphere swirls around the globe with a force that can be as delicate as a soft kiss to the cheek or strong enough to throw cows into the air and rip homes off their foundations. It is a tremendous global resource.

Immigrants from Europe sailed to North America on the power of the trade winds, and as settlers moved west establishing farms, thousands of the ubiquitous multi-bladed windmills towered over barns and houses, drawing water from wells, and, later in the 1900s, running small electric generators that lit up homes. Hundreds of thousands of these wind dynamos powered farms after the Second World War right up to the mid-twentieth century before power lines reached rural areas in the 1950s.

Wind technology was largely abandoned when electrification spread to rural areas, although there was a resurgence during the oil crisis of the 1970s to offset rising energy costs. Today, wind is back in favour around the world on a much larger scale to tackle the more serious issue of climate change.

The wind carries more energy than we can use. It is abundant, powerful, and free, but like solar energy, is spread out over vast areas, and it doesn't always blow, making it a challenge to capture. Still, worldwide, wind energy production has doubled since 2015, totalling 743 gigawatts by 2021 with more than half of that in China.[1] That represents less than 7 percent of the world's total energy production, but some countries are turning more to wind power: in Denmark, it provides almost half of the electricity supply, and in Ireland, more than one-quarter.[2]

The trend is expected to continue as the cost of wind technology comes down while fossil fuel prices go up.

How a Wind Turbine Works

Wind turbines are simple in principle. Think of an electric fan running in reverse. Rather than an electric motor spinning the fan blades to make wind to cool a room, a turbine has the wind blow against the fan blades, causing them to spin, and the electric motor becomes a generator that makes electricity. There are two basic approaches.

The most common is called a geared system. A typical wind turbine rotor turns quite slowly, around 18 revolutions per minute, which is one turn every three seconds. A shaft from the rotor connects to a huge gearbox in the nacelle at the top of the tower. It acts like the transmission in a vehicle, using a system of gears to increase the rotational speed, in this case by a factor of one hundred. The output shaft rotates at 1,800 rpm and is used to spin the generator. To keep the power output constant, these turbines are controlled so that they run at a constant speed no matter how fast or slow the wind is blowing.

A simpler design, known as direct drive, eliminates the heavy complex gearbox and basically turns the rotor into one giant generator. The central hub turns a cluster of magnets that rotate inside a huge ring of electric coils 10 metres across. With fewer moving parts, this one large, slow-speed generator concept is appearing in some of the biggest wind turbines.

The only issue is if you want to make a lot of electricity. For that, you need to catch a lot of air, which means scaling the whole thing up to fantastic proportions.

Early windmills had rectangular, flat-sided blades that acted like paddles, deflecting the air to make them spin. In aircraft terms, this is the angle of attack, or the angle of the wing facing the wind. During takeoff, when the aircraft speed is low, the leading edge of the wings

The Dutch became masters of wind technology to pump water out of low-lying land and into rivers beyond dykes.

Image by Pexels by Pixabay

is tilted up, increasing the angle of attack so that more air is deflected downwards, increasing lift. Once the plane reaches a higher flight speed, the wings lower their angle of attack by levelling out because the air exerts more force on the wing the faster it moves.

While flat blades on a windmill are effective in deflecting the wind, they are not the most efficient aerodynamic shape. As the air swirls around the backside of the blade, turbulence can form, which increases drag. Modern turbine blades are more like aircraft wings with a sculptured form that allows the air to flow over their surfaces in a very smooth manner and develop more lift. The upper surface of aircraft wings is curved. As the air passes over the curved surface, it speeds up through a principle called the Bernoulli effect. Back in 1738, Daniel Bernoulli discovered that when air speeds up, pressure decreases. So, the fast-moving air over the top of a wing causes low pressure that draws the wing upwards. That extra lift, combined with the flat underside of the wing that uses angle of attack, enables even the largest aircraft to lift themselves into the sky.

The blades of a wind turbine use the same principles to actually fly through the air. In fact, the windmill blade has evolved out of aviation.

According to Dr. Katherine Dykes, senior researcher at ETU Wind

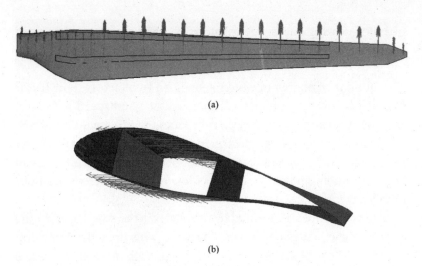

(a)

(b)

Typical cross-section of a wind turbine blade showing direction of motion.
This shape is similar to an aircraft wing.

Wind Energy Science, European Academy of Wind Energy

Energy in Denmark, changes in blade design were driven by aviation
research in the twentieth century.

It was after World War One, following on all the developments
that were happening in aviation and aerospace, all the research
around propellers, airplanes, that led to the first wind electric
machines. A lot of the first commercial machines came out of
that post-war period.

Most of them had three blades, horizontal axis, some had
two blades. It all came out of the science around propellers
and aerodynamics that allowed people to design turbines that
incorporated lift to efficiently generate electricity. Thousands
of these three- and two-bladed machines were installed all over
America and Europe until massive rural electrification came
in and took over in mid-century.

So, we have a history of two- and three-bladed machines.
It wasn't one or the other; both were widespread.

Modern wind turbine blades are considered wings, with thick
roots at the hub that taper to thin tips and curved cross-sections
that enable them to fly through the air and develop lift. A wind

turbine is a wing on its side with the lift force directed sideways in the direction of the spin.

The amount of energy in wind depends on two factors: how fast it blows and how much of it you can catch. As wind speed increases, the energy it carries goes up by a cube factor, which means that if the speed doubles, the energy goes up by eight. That's a lot, so you don't need huge wind speeds to capture a lot of energy. Most turbines can begin producing power in winds as low as 12 kilometres per hour (7 miles per hour).

The amount of air captured by the blades is the swept area, or the big circle the blades cover as they rotate. This area also rises dramatically as the blades get longer. Think of each blade as the radius of a circle, so to determine the swept area, just apply the trusty "pi r squared" formula you learned in school, but probably seldom use, that calculates the area of a circle. (Remember, pi are round, cake are square.) The r squared part of that formula means that if the radius increases, or the blades get longer, the area goes up by the square of that increase. In other words, double the length of the blades, the swept area goes up by four. That's one reason turbines keep getting larger and larger, to increase the swept area and capture as much wind as possible.

The challenge for wind energy is finding areas where wind is constant, with enough room for the machines to be as large as possible. That's not always easy around cities.

It is hard to get a sense of how big wind turbines really are when you see them out in a field. They have to be separated from each other, kept away from power lines, and at a distance from homes, so there is nothing to compare them to for scale. But modern turbines are truly enormous.

While they come in many different sizes, the average wind turbines you see on land, generating around 2 megawatts each, are made with 70-metre-long blades; each one is longer than the entire wingspan of a 747 jumbo jet. An average person would take about two minutes to walk from the base of a single blade out to the tip. There are three of them on a rotor, so the total diameter from one blade tip to another could be one and a half times the length of a football field.

Those three gigantic blades must be held up by even taller towers measuring 90 metres, so when a blade reaches the top of its arc, the tip is 160 metres above the ground. That is roughly the height of a fifty-storey building, or half the height of the Eiffel Tower, only 9 metres below the top of the Washington Monument, taller than the Great Pyramid of Giza in Egypt, 50 metres higher than the Saturn V moon rocket, and more than one and a half times higher than London's Big Ben or the Peace Tower on Canada's Parliament Hill.

That's an average wind turbine. At time of writing, the world's largest is the GE Haliade-X. Intended for offshore use, it reaches 260 metres above the sea.[3] That's the height of an eighty-storey building with a swept area of 18,000 square metres, or seven American football fields.

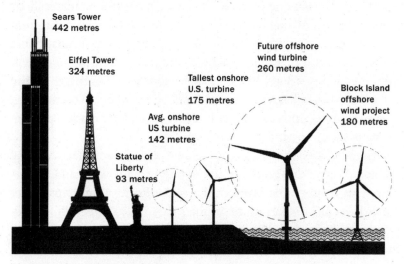

The scale of 15 MW wind turbines.

Courtesy of Bumper DeJesus, Princeton University

These giant machines put out an impressive amount of power. At the time of writing, the GE Haliade-X generates 12 megawatts, which provides enough electricity to power sixteen thousand European homes.[4] Two other companies, Vestas and Siemens, are offering units putting out 15 megawatts. Just one turn of the blades on these monsters can run a single home for two days.[5]

Unlike tall buildings that stand strong through their weight (although some of the tallest still move a little in strong winds), building these turbine superstructures is a challenge. They are moving, dynamic machines handling powerful wind forces.

The super-long wind turbine blades must be incredibly strong just to hold their own weight, and again, they took a lesson from the aircraft industry. When the largest airliners, the Boeing 747 or Airbus A380, are on the ground, their enormous wings must extend straight out from the body with two giant engines hanging underneath. (Try holding a bucket of water in your hand with your arm straight out to the side and see how long you last.) When you consider that the wings are attached to the plane only by their roots, that takes tremendous strength.

Turbine blades, which are even longer than aircraft wings, are mostly made from one piece of fibreglass, a composite material, so that they will be strong enough to not only hold themselves out from the hub but also withstand the force of the air pushing against them. However, because they are made in one piece, transporting them from the factory where they are made—such as in Denmark—to the site where they will stand—which could be anywhere in the world—presents interesting challenges. In many cases, local blade manufacturing is set up to get around the problem.

When I visited the TransAlta wind farm site in southern Alberta, Canada, Wayne Oliver, operation supervisor for western Canada, toured me through the vast foothills of the Rocky Mountains where more than one hundred white turbines stand like giant trillium flowers across the land, all turning serenely. The company operates more than nine hundred turbines across the province.[6]

Cowley Ridge was the first installation in 1993 with old lattice towers and multi-blade rotors that were 300 kilowatts each.

In 2001 they were 660 kilowatts, 2002, 1.2 megawatts, 2020 4.8 megawatts. In the future they will be up to 10 megawatts.

The challenge is getting turbines that large to remote locations where the wind is. Do you have cranes big enough to assemble them? Southwestern Alberta is easy because it's flat and downwind of the Rocky mountains.[7]

In mountainous areas, some blades are too long to travel around curves and through tunnels by train, so they must be moved by specialized truck transport, often at night. Huge cranes, brought in sections and assembled on-site beforehand, must be capable of lifting many tons of tower components, generators, and blades to incredible heights. But all that height pays off. Undisturbed by trees and buildings, winds are stronger and steadier the higher off the ground you go. The size also pays off. The power output of one giant machine can replace ten from the past. But building monumental structures takes a monumental effort.

> The tower weighs 300,000 pounds; the nacelle, which contains the gearbox and generator is 150,000; the hub 45,000; and 15,000 per blade for half a million pounds total. Components come from Denmark, Germany, and China, and are transported by ship, rail, and truck.
>
> Huge 600-ton cranes need to be assembled on-site, requiring twenty-four semi trucks full of parts just for cranes. It takes two cranes to pick up each section. The first section of the base must line up on 144 bolts on the foundation. And there is always one that doesn't line up.
> *Wayne Oliver*

One company that is getting around the problem of giant cranes is Enercon, which uses a self-climbing crane. It hugs the tower like a bear climbing a tree, lifting itself higher as the tower grows, eliminating the need for the gigantic cranes on the ground that need to reach the enormous heights of the modern turbines.

How large can these giants get?

> The question of turbine size comes up a lot. I've not seen any analysis saying this is the absolute limit. I remember when we talked about 5-megawatt turbines as being huge and 10-megawatt being outlandish and not a possibility. There was a time when 5 megawatts was considered outlandish. The turbines in the 1980s during the wind rush in California were 55-kilowatt machines. We're talking one hundred or more times increases in size.

We're now realistically talking about 20-megawatt turbines.
There is research out there considering 50-megawatt machines,
at least from a conceptual level. If you were to ask me today,
I would say 50 megawatts sounds outlandish, 20 megawatt is
doable considering where we are, with 15-megawatt machines
on the horizon. But every time someone says, we'll never build
a turbine that big, they've been wrong. So I'd hate to be quoted
on that, then be shown wrong.
Dr. Katherine Dykes

The other factor in wind energy is the force of the wind itself, which,
as any sailor knows, becomes stronger the faster the wind blows.
That's great news for areas where wind speeds are commonly high,
but you don't want the blades to spin out of control during a gale.
The giant blades would be ripped from their roots and the whole
structure would shake itself apart, which has happened in the past.
Older turbines are designed to run at one speed no matter how fast
or slow the wind blows. This not only ensures that the spin does
not get out of control but also synchronizes the frequency of the
wind turbine output with that of the electrical grid. If you think
about it, no matter where you go, you can plug your device into a
wall socket and the frequency and voltage of the electric current is
always the same. In North America that's 110 volts at 60 cycles per
second. The entire electrical grid must be synchronized this way so
that all our electrical equipment can function. Wind turbines have
to fit into that pattern, which is why a field of turbines can look like
a choreographed dance where they all turn together in a synchronized chorus line.

More-modern wind turbines can get around this problem and
operate at different speeds depending on the wind, but they require
electric converters that can accept the variable speed of the rotor,
rectify it to DC power, and then invert that to the grid frequency.
Some of the largest machines operate this way, making them more
efficient, but they still have an upper speed limit.

A second reason to control turbine speed is for sound. Gigantic
machines may not appear to be turning very fast, but the blades are
now so long that the tips sweep a huge arc through the air in only a

Serrated wind turbine blade designed to imitate an owl's wing for quieter performance.
Image courtesy of SMART BLADE, GmvH, Germany

few seconds—meaning the outer edges of the blades are travelling over 100 kilometres per hour. Just a slight increase in rotor speed can bring the blade tips close to the speed where compression waves form, generating noise. A lot of research has gone into the design of blade shapes to minimize noise, including experiments that incorporate "feathering" along the trailing edges, a technique borrowed from nature.

If you have ever seen an owl in flight, you probably didn't hear it because the predatory birds are one of the few species who are totally silent fliers. Often called cats with wings, their huge eyes and large wings give them the ability to sneak up on their prey in full stealth mode. Scientists have wondered how they manage to fly without making the flapping sounds of pigeons and other birds.

Close examination of an owl's wing shows a line of specialized feathers along the leading edge sticking up like the teeth of a comb that have been bent backwards. It is believed that these structures break up the airstream into tiny vortices or whirlpools that are absorbed by soft fluffy feathers along the top of the wing. This creates a layer of air that clings to the wing so the air passing over the wing "sees" other air (as engineers say) rather than the surface of the wing itself. The result is less sound. Another set of comb-like feathers running along the trailing edge of the wing allows the air to leave the wing with less turbulence, reducing noise further. It works for owls, so modern windmills are borrowing a lesson from nature by adding comb-like structures to windmill blades resulting in a 40 percent reduction in noise.

With speed limits on wind turbines because of structural loads and noise, the only way to get more power out of the wind is to go big. It comes down to the amount of force acting on the blades that is turned into torque, or the twisting force applied to the central shaft. Longer blades are catching more air because of their increased swept area so the torque they produce is much higher. That extra torque can drive a larger generator and deliver more electricity. A large airliner and a small business jet both fly at around 850 kilometres per hour, just below the speed of sound. But the larger wings of the airliner allow it to carry much more weight because they catch more air and develop more lift.

Standing directly under one of these giant spinning turbines is a surreal experience where it is difficult to grasp the scale.

The gleaming white tower reaches straight up to touch the sky. Blade tips swing by every second or so with a gentle hiss. In fact, if engineers hear extra noise, that indicates to them that something is wrong, perhaps a flying object hit a blade and chipped out a piece or, sadly, in some parts of the country, bullet holes are disrupting the airflow. Looking straight up at the spinning rotor from the base of the tower, you can see how the blades are curved backwards by the force of the wind, like the petals of a flower in bloom. It is amazing how structures that long can bend so much without snapping off, but that is part of the design. The nacelle at the top is angled back 7 degrees to allow the blades to bend without striking the tower. More-modern blades are designed with a built-in bend as well as the ability to twist in the wind to reduce loads on the structure. If the blades didn't bend, they would likely break.

You may have had the uncomfortable experience of looking out the window of an airliner when it hits turbulence and seeing the wings flapping up and down like that of a bird. It might seem like they are about to snap off, but they are designed for that movement. If they didn't bend, they would likely break. All that flexing distributes the stresses along the entire length of the structure. Aircraft wings, or wind turbine blades, could be made stiffer, but they would need extra reinforcement, like the girders of a bridge, which would add weight—not good for structures meant to fly through the air.

The more important property of flexible structures is how they dissipate energy. A solid wing would focus all the forces acting on it down to the root, where it attaches to the plane body or the wind turbine hub. Any weak points at the base will become weaker as those forces act on them, which could lead to failure. But if a wing or blade is allowed to bend, the forces are passed along the entire length and dissipated evenly. The bending is like a wave that travels from the root out to the tip carrying energy with it.

Turbine blades are manufactured mostly in one piece using tremendously long moulds. Carbon fibre mats, resins, and other compounds are laid out with extreme care so that the blade emerges as a single flawless shape, 45 or more metres in length, that is capable of flexing without breaking. To make transportation easier, the largest blades do have tip sections that are added on-site.

From the outside, the ponderous motion of the giant rotors seems almost serene. But there is much more going on than meets the eye.

Towers can move as much as eighteen inches off centre in ninety-kilometre winds, so thirty-six inches total sway, or one metre back and forth. Some workers get seasick up there.
Wayne Oliver

As wind speed increases or decreases, the blades change their angle of attack to the wind. They face more broadside at low wind speed and turn more edgewise as the wind increases. Windmill blades are constantly adjusting their angle of attack until the speeds rise to about 80 kilometres per hour. Above that, the blades turn completely edgewise to the wind and rotation stops to prevent damage to the structure. Modern turbines are equipped with sensors and self-diagnosis to shut themselves down if conditions become too dangerous and there's a potential for damage. As I found out during that Florida storm, you can get too much of a good thing, and sometimes you have to heave to in a storm.

On the other hand, those ocean winds are an even more valuable resource, which is why many wind farms, involving the largest turbines on the planet, are being built offshore.

> Offshore, you typically have a higher wind resource, especially more wind closer to the surface so you don't have to build the machines as high. Power for a wind turbine is a function of the velocity cubed, so going offshore is better because of the resource, more wind.
>
> Another reason to go offshore, and why Europe has done that a lot, is because onshore, especially along the coastal regions where there is a lot of demand for electricity, there is not much land on which to build wind turbines. So going offshore allows you to access high winds near load centres.
>
> In the US, they build a lot of wind turbines in the interior, but they need long transmission lines to get to the coast where the load is. But just off the coast of California and the Eastern seaboard of the US, there is a lot of high-wind potential. If you can make offshore wind economically feasible, you can put the wind where it is needed near all those big cities.
>
> *Dr. Katherine Dykes*

Wind turbines have evolved over time from the multi-bladed fan shapes on farms to the three-bladed propeller, which seems to be the most efficient design. Experiments with alternative designs have had mixed results. The vertical axis turbine involves a large central axle standing on end with the blades bowing out from the top in graceful curves that connect at the bottom. They're known as eggbeaters. Some variations use corkscrew shapes or vertical wings, but the big advantage to this design is that the wind can approach from any angle and the blades will still spin. The propeller style must be constantly steered directly into the wind.

> There was strong support for and research into vertical axis machines in France, in Canada and the US, several commercial machines were built, then two-bladed machines and three-bladed. The three-bladed horizontal axis machines we see today

were more reliable and more robust, which made them more acceptable overall, which is why they became the platform standard for the industry.

Vertical axis have some really nasty loading in turbulent air at the base of the tower, which is a difficult engineering problem to solve. Two-bladed have some nasty gyroscopic loadings, especially in yaw, that creates some additional engineering challenges that three-bladed machines don't have. And while those engineering challenges can be resolved, we now have decades of experience with the three-bladed design, which has made them very low cost, highly efficient, and reliable machines.

Dr. Katherine Dykes

Vertical axis turbines have found many uses in smaller applications, such as on rooftops, but when scaled up to industrial size, problems emerge. According to Dr. Michael Ross, NSERC Industrial Research Chair in Northern Energy Innovation at the Yukon Research Centre, Yukon College:

> Part of the issue is the area that is swept. You have a smaller area, but also, the wind is stronger the higher you go above ground. So for a vertical axis, you have to build it a lot taller and more infrastructure is required for not as much swept area as you can achieve with a horizontal axis wind turbine.[8]

The taller you go, the heavier that structure becomes, and all that weight rests on one single main bearing at the base that still has to turn smoothly. That's a lot to ask of a bearing. Any vibrations can do significant damage, which is why the vertical axis turbine has fallen out of favour on a large scale, but has some applications where space is limited.

Wind is very visceral; people feel it on their face and think it is windy. But actually, the average wind speed in many urban locations is quite low. With rooftop wind, you are in a very turbulent wind environment so getting a wind turbine to produce

a reasonable amount of energy is hard. The capacity is an order of magnitude smaller than what a wind farm could produce.
Dr. Curran Crawford, Institute for Integrated Energy Systems, University of Victoria

But why only three blades on the large turbines? Why not just two, or four or more, like the traditional multi-blade farm design from the past? Wouldn't that make more power?

It's not so much that more blades will give you more power. Let's say you just have one blade. It will be unbalanced, it will be heavy, and it probably won't spin. Try adding a second blade, you'll be able to physically balance the turbine, but you would need to have higher rotational speeds or longer blades to match the potential power output that can be achieved with a three-bladed design.

If you have three blades, connected 120 degrees apart from each other, just like a tripod, you have a balanced system that produces more power than two blades.

Part of the issue of having even more blades is that you might think that you get more power, but given the added weight and cost, the blades don't actually extract significantly more power, and the pressure differential on the different sides of the wind turbine and the area that it encircles will still be the same. The three-bladed design is the sweet spot for performance versus cost.
Michael Ross

It really comes down to economics. A two-bladed turbine with blades the same length as a three-bladed would have to spin 22 percent faster to have the same output. That puts more stress on the structure and, more importantly, increases noise. Adding more blades will add more to the cost of manufacture, as well as increase the weight and complexity of the structure, so it's not worth the effort. Also, a four-bladed hub would have to be square, which is not as strong as the circular three-bladed centre. That's why the three long blades have turned out to be the most common.

No matter how many blades are involved, all wind turbines have a fundamental limit called Betz's law, developed by German scientist Albert Betz in 1919. This law states that no more than 59 percent of the energy in the wind can be extracted.[9] That is because the wind must pass completely over the blades then keep going past them so that the flow keeps moving. If all the energy was removed from the wind, the air would be stopped, which is otherwise known as a wall. And walls don't move much. The only way to get more energy from wind is to make the turbine larger, which is why modern machines have reached such gigantic proportions.

Most turbines operate around 50 percent efficiency, which is close to the Betz limit. That means the air flowing out on the downwind side is more diffuse and moving more slowly than the air coming in. Turbines are spread apart from each other to ensure that none are in the shadow of another.

Wind energy, as a resource, is not evenly distributed around the world. Like oil fields, which are only found in certain underground reservoirs in places such as Texas, Alberta, and the Middle East, wind blows more consistently in some places than others. The foothills of the Rocky Mountains are a prime location since winds blow down from the mountains on their way across the plains. Coastal areas with prevailing onshore winds are another attractive location; and if there are shallow seas, offshore installations will capture consistent sea breezes. European countries are exploring the latter option, which allows turbines to be much larger since the huge parts can all be delivered by ship. However, construction and maintenance costs over water go up considerably.

One interesting challenge for wind power is in the Far North, where communities are small and separated by vast distances. Many are only accessible by air. These communities are powered by diesel generators, which not only is expensive but also releases a lot of carbon into the atmosphere.

We emit the most greenhouse gas per capita than any other place in Canada. It is a challenge in the north because we have lower population density than in the Sahara Desert.

In terms of meeting the conditions of the harsh winter and remote environment, risk is a high consideration. But renewable energy has the potential to reduce the dependency on fossil fuels.
Michael Ross

In the North, everything is expensive, mostly because of the cost to get anything there by air. Because of the extra cost, and the harsh conditions, replacing the diesel generators with clean wind technology presents a challenge.

There is a balance between risk mitigation and viability. For example, even if there is a leading-edge technology, if it gets implemented into a fly-in community and something goes wrong or something needs to be fixed, someone has to be flown in to fix it, or it has to be flown out, even at the end of the project's life cycle.

The economies of scale aren't there, and you have a high risk that something could go wrong because it is an extreme environment and it is isolated. A power outage in the South is often an inconvenience; in the North it can be critical.

We don't want to see a graveyard of leading-edge technology just because it was too much of a risk to implement, or too costly to repair.
Michael Ross

Another issue in the Far North is ice accumulating on the blades, which reduces their aerodynamic efficiency and adds dangerous weight. Heaters can be added to keep the blades frost-free, but that requires tapping into some of the energy generated by the turbine that would otherwise go to homes.

Nothing is easy in the North.

A field of gleaming-white, towering wind turbines all rotating together like a huge aerial ballet may be a symbol of our green energy future. But to many, they are an eyesore and literally a headache. People living close to industrial turbines have complained of noise that continues through the night, ground currents, and light flicker. They report symptoms of sleep deprivation, fatigue, headaches, heart palpitations, anxiety, and depression. As a result, in Canada, wind turbines must be at least 5 kilometres from residential areas but no more than 74 kilometres from existing power lines.[10] Too far away and costs increase due to transmission losses over long distances.

Conservationists have chimed in with concerns over the impact of large wind turbines on non-human residents, mainly birds and bats. You might think that a large structure standing more than 100 metres tall that appears to be rotating rather slowly would be easily avoided by flying animals. This may be true during the day, but many bird species migrate at night, and bats are almost exclusively night fliers. Furthermore, the animals don't have to actually be struck by a spinning blade to be killed. The high speed of long turbine blade tips moving through the air creates swirling vortices and pressure differences that can rupture the lungs of a bird or bat that flies too close, knocking them out even though they haven't touched the blade.

Sadly, wind farms have become yet another in the long list of human obstacles to birds and bats, with an estimated 140,000 to 328,000 birds killed every year in the US alone.[11] That is a terrible number, but it is less than the millions of birds killed by flying into tall buildings or power lines, or being caught by domestic cats. Efforts to reduce the impact of wind farms on wildlife are considered when deciding the location of the farms and their operation. Situating wind farms away from known migration routes and sensitive habitats is the first step. Another is to either shut the turbines down when migrating species are in the area or increase the wind speed at which the turbines start up since animals tend not to fly as much in higher wind speeds.

We did a study with the University of Calgary that found we were killing about two hundred bats a year. We changed the start-up speed during migration time and cut mortality in half.
Wayne Oliver

Deterrents are also used. Ultrasonic generators emit sounds that humans cannot hear but that are uncomfortable to bats. Painting one of the three blades a dark colour makes the machine stand out more, and ultraviolet light shone on the blades makes them more visible to animals at night.

Wind energy has huge potential, but incorporating it into our energy mix is a challenge. The power output is literally at the whim of the wind. Some local residents may oppose installations while, on the other hand, farmers earn extra income by hosting wind turbines on their land, where they can still farm and ranch among the towers. When visiting the turbines at the Pincher Creek site, we had to slowly push our way through herds of cattle who not only didn't seem to mind the presence of the big machines but also used the bases of the towers as rubbing posts. For the ranchers and farmers, it is steady extra income from leasing their land. It is important that wind farm operators work with local farmers, ranchers, communities, and conservation authorities to minimize the impact of wind energy on wildlife and the environment.

Every form of energy conversion comes with costs and consequences. It comes down to how much we are willing to pay or put up with. But we have to always keep in mind that every kilowatt produced by wind is power that was not generated by burning fossil fuels.

> There is a possibility of wind really becoming a backbone foundational technology in the future. We see wind and solar as being workhorse technologies for our future clean energy system. It is definitely technically possible. There is certainly work to do on other sectors, the industrial side, transport, and all these other things, but there is certainly the potential for wind and solar to provide the majority of our energy needs of the future.
> *Dr. Katherine Dykes*

Of course, wind turbines are not carbon free. Fossil fuels are used during manufacture, transportation, and construction of the giant machines. But that carbon footprint is tiny compared to burning fossil fuels directly.

According to the National Renewable Energy Laboratory in the United States, 1 kilowatt hour of electricity from wind energy produces 11 grams of carbon dioxide (this includes emissions from the manufacture of the turbines). By comparison, a kilowatt hour from natural gas produces 465 grams, and from coal, a whopping 980 grams of CO_2.[12] In other words, the carbon footprint of coal is ninety times greater than wind, and natural gas forty-five times greater. These climate-changing fossil fuel emissions are at the core of the case for wind power. It is the long view that counts if we are to meet our Paris Agreement and keep the climate from warming above the 2°C threshold.

Sailors have been capturing the wind for centuries for transport and sport. And it continues to this day. Every year, millions around the world watch the America's Cup yacht race where international teams sail high-tech crafts that seem to fly above the waves suspended on wing-shaped hydrofoils that enable them to travel faster than the wind itself. Even large cargo ships are flying giant kites off their bows that pull them across oceans, saving 10 tons of oil a day thanks to the power of the wind.

For me, facing the thunder squall off the coast of Florida, I struggled to furl in the jib sail so the wind had less surface to push against. The boat righted herself, giving the rudder more purchase in the water and allowing me to keep the bow pointed into the mountainous waves. I was in survivor mode doing whatever was necessary to avoid the rolling whitecaps and keep the boat from being swamped. It was a wild and scary ride.

Then, just as suddenly as it arrived, the storm departed. The winds dropped off immediately, returning the day to typical Florida sunshine as though nothing had happened. It was the scariest twenty minutes of my life. Later, during that same trip, when a squall line appeared on the horizon, I prepared the boat ahead of time by shortening the sails, harnessing myself to the boat, and battening down the hatches, and then rode out the storms under control. But even though I learned to ride storms, it gave me tremendous respect for the power of the wind and the sea.

The Earth's atmosphere will be in motion for billions of years. It carries tremendous amounts of energy. Our challenge, as with the sun, is to find innovative ways to capture it.

Chapter Four

Ocean Wave Power

Surfing the energy out of waves.

"Wait until the water fills the gap, then jump in and swim out right away. When you come back, let the wave lift you up, grab onto the rock, wait until the water washes, then you can climb out."

Those were the instructions given to me by a local resident of the Big Island of Hawaii as I was about to go snorkelling in the clear blue waters of the Pacific Ocean. This wasn't a beach, but rather a sheer drop where black lava had flowed down the mountain to the sea, broken off, and tumbled into the depths. The rock face was vertical and very jagged, but there was one spot where a long crack a few metres wide ran in from the ocean. In a steady rhythm, the chasm filled with water every ten or fifteen seconds as waves pushed against the shore, then emptied in an outward flow like a dam had burst at the mouth. It was a dance between land and sea that has been going on for billions of years and I was about to join in.

To enter this maelstrom, I had to time my leap perfectly or I would be smashed against the jagged walls and ripped to pieces by the powerful surge. The local resident told me they swim here all the time, so to just go for it. Fortunately, I have been a swimmer all my life so I'm not afraid of water. But this was something entirely different from the tranquility of a Canadian lake. This water was alive.

Donning a mask and snorkel, I waited for the water to pause at its highest point then took the plunge. Instantly, the warm salty water carried me out through the gap. I swam hard with the current so I would get far enough away from the cliff face to avoid being drawn back in with the next wave.

A ways offshore, everything calmed down to a gentle bobbing motion as the deep-blue waters of the Pacific rolled under me.

On the first dive under the surface, I faced a sheer wall of rock that dropped steeply down into the depths with no bottom in sight. The Hawaiian Islands are a chain of the world's largest volcanoes that rise straight up from the floor of the Pacific. Measured from its base on the sea floor, 5,486 metres below sea level, to the summit of Mauna Kea, 4,206 metres above, the Big Island is taller than Mount Everest. I felt like a small bird hovering above the flanks of the largest mountain on the planet.

These islands stand alone in the middle of the Pacific Ocean with no barrier reefs surrounding them, like those that shield islands in the Caribbean from big deep ocean waves. This is why Hawaii is one of the surfing capitals of the world.

The rock wall had a greyish colour from all the marine life that had grown over it, while hundreds of colourful fish pecked away at it for food. I struggled to hold my position against a strong current that was flowing back and forth along the rock face. Then I realized the fish were not swimming against it but drifting with it, several metres in each direction. I relaxed and allowed myself to drift along with them, only swimming when the water was moving in the direction I wanted to go.

A short distance ahead lay a huge boulder the size of a single-car garage; it had obviously tumbled down from above and was leaning against the wall. A gap near the bottom formed a tunnel large enough to swim through. Fish tend to gather in protected areas like that, and I could see clear water on the other side. I'm pretty good at underwater swimming and figured I could easily make it through on one breath.

(Two notes of caution here. I was violating a cardinal rule of diving: never swim alone. An underwater mask distorts light so distances are deceiving. I can only use the excuse that I was young and foolish at the time.)

As the surge began to move towards the gap, I dove down and followed it through. Sure enough, the passage was chock full of fish, both on the walls and ceiling of the tunnel. I was in the middle of a natural aquarium. The opening on the other side was tantalizingly close as I began to rise to the surface, but suddenly my movement was halted as the water reversed direction sucking me back in. I was not wearing fins on my feet so I couldn't produce enough thrust to push against it, and it was too far in to go back the way I came. All I could do was struggle to remain in one place until the next surge came through to carry me out.

It is an odd feeling being trapped underwater and realizing you are not going to be breathing again as soon as you expected. Fortunately, I'd had a lot of experience remaining underwater for minutes at a time, so I didn't panic—but by the time the next surge arrived and I burst to the surface with a huge gasp, I realized I had seriously underestimated the power of deep ocean waves.

According to a study by the University of Victoria on Vancouver Island, a survey using ocean buoys showed that the power of ocean waves is a global resource that has a higher concentration than both wind and solar energy. Rough estimates of global available wave power are 1 terawatt near shore and 10 terawatts offshore.[1]

Despite the great potential, capturing that energy and turning it into electricity is a very difficult task. No commercial technology is in operation, largely due to the financial challenges of providing reliable large-scale electricity generation. The ocean is a hostile environment so any technology designed to remain there is expensive to operate and maintain.

Waves are, in fact, a form of wind energy because all waves are generated by air passing over the surface of the water. The wind folds the water like a rumpled carpet into crests and troughs. The waves move in the same direction as the wind and their heights are determined by the wind speed. On the ocean, where there are long stretches of open water with no obstacles, waves can become very long as well as very high. The only exception are tsunamis, which are produced by earthquakes on the ocean floor; while they carry tremendous energy, their occurrence and location is too unpredictable to be considered for wave energy.

Like all energy resources, wave energy varies from location to location. For waves to grow to sufficient size, they need a long stretch of open water where steady winds can build them up. That means along coastlines where the prevailing winds blow onshore, such as the west coasts of North America and Europe.

Generally, wave energy is highest at higher latitudes, above 30 degrees north or south, which is also where solar is the least available. The tropics also have great resources, but there are different characteristics of the waves, but good results still.
Dr. Bryson Robertson, director, Pacific Marine Energy Center, Oregon State University

Water waves are one of the few places where you can see energy in motion. Other forms of energy, such as heat or electricity, are invisible; even a light beam appears unchanging because you cannot see the photons moving along at the speed of light. But the parade of waves across the surface of water is the movement of energy itself, energy that has been captured from the wind, made by the bow of a passing ship, or generated by a simple stone thrown into a pond. The interesting part is that the water itself is not actually moving that much, only the energy flowing through it.

Despite what you see standing on a beach watching waves roll towards you, the water is not heading your way. Sure, whitecaps can be pushed along the front side of a wave when it becomes too tall and steep, but most of the water does not move ahead. Picture sports fans seated in stadium rows performing a wave by standing and raising their hands in sequence: each individual only moves up and down in their seat while the wave passes through the crowd. If water did move along with the waves, then it would keep piling up on beaches

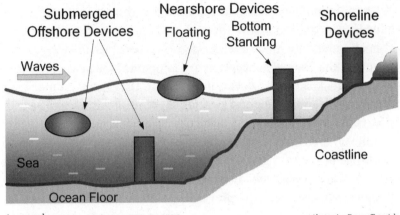

Assorted ways to capture wave energy.   Alternative Energy Tutorials

until every beach became flooded. But that doesn't happen. Even the biggest surfing waves stop after they hit the beach and then retreat back to the sea.

Capturing wave energy and turning it into electricity is not simple because the water moves in different ways depending on its position within the wave. As a wave passes one spot, there are four different motions the water goes through. First, there is the obvious up and down undulation at the surface of the water, known as heaving (the action that filled and emptied the gap I jumped into on the rocky Hawaiian shore). Just below the surface, the water goes through a circular motion like a log rolling back and forth, called pitching (the elliptical action the fish and my body were doing along the rock face). Then, if the water is shallow, it will surge back and forth along the bottom (the action that almost prevented me from getting through the tunnel). Finally, there is the up and down motion of the air directly above the water surface. This means there are several different approaches to harnessing wave energy.

Over decades of research, devices of many types have tried to capture one or several of these cyclic motions and use them to drive turbines that generate electricity. Unlike wind power, where the technology has converged into the three-bladed propeller style, or solar energy, which has settled on photovoltaic, there is no one design that has proven to be the best for wave energy.

There are different technologies that can take advantage of the wave periods, and technologies to take advantage of the wave heights. We have devices that will be mounted to the sea floor, and as the wave goes by, they will raise and lower. And they will take advantage of the potential energy there.

We have devices that take advantage of the orbital velocity within the wave. As the wave passes, individual particles spin, and we have devices that take advantage of that rotational kinetic energy.

We also have devices that take advantage of the pressure wave that passes underwater along the sea floor as the wave propagates on the surface. So there are a whole slew of different technologies.

Dr. Bryson Robertson

Research continues. So, let's look at the various technologies starting from the shoreline and working out to sea.

Voith Wave Generator

One successful system, designed by the German company Voith, is also one of the simplest. The first test installation was along a rocky shoreline in Scotland in 2000. It is a nondescript rectangular concrete structure resembling a Second World War bunker. The lower section extends down into the water and is open at the bottom. As waves enter the chamber, air inside is pushed up and squeezed through a narrow opening under pressure into a second upper chamber. That high-pressure air spins a turbine to generate electricity. As the wave retreats, the air in the upper chamber is sucked back down through the same opening into the lower chamber and spins the turbine again. The turbine is designed to always turn in the same direction no matter which way the air is flowing. The beauty and advantage of this system is that it involves only one moving part powered only by air, which means there is no contact with the sea water so no issues with corrosion.

After ten years of successful operation, an improved version involving six turbines was incorporated into a seawall in the port of

Mutriku on the Basque coast of Spain in 2011. That provided 300 kilo-watts of power to 250 homes in the town.[2]

Unfortunately, Voith shut down its operations in Scotland due to a lack of investors, even though the company showed that the technology works. This is a common theme in wave energy start-ups.

Wavestar Project

Wavestar, a project in Denmark, resembles a giant centipede: twenty paddles, each with a 10-metre float on the end, stick out the sides of a long, fixed structure. Passing waves move the paddles up and down, which act as pumps to drive hydraulic fluid into reservoirs under pressure. The fluid is used to run hydraulic motors that run generators. The advantage of this machine is that the paddles operate independently, moving in sequence as a wave passes through, so one of them is always in motion, which reduces the cyclic nature of waves. So far, scale models, up to half-size, have been tested at sea proving the concept.

At full scale, each unit can produce 6 megawatts of power, enough for four thousand homes. The structures stand on the sea bottom in shallow water. Alternatively, the company Eco Wave Power in Israel has mounted the paddles on the outside of a breakwall to

Centipede-like structure of the Wavestar generator.   Photo courtesy of Wave Star A/S

catch the energy of surf hitting the shore. The advantage to this system is that all the parts are above the waterline, making it easier to service the components.

Other Designs

Finland-based AW-Energy is developing what they call the WaveRoller, a 350-kilowatt device that is basically a giant paddle sitting upright on the sea floor. As the waves pass over, the surging water causes the paddle to flap back and forth like a big fan, driving hydraulic pumps. The device sits in 10 to 15 metres of water so it is not subject to the pounding of storm waves rolling across the surface. The company is planning to scale up to 1-megawatt units that can be deployed anywhere in the world.

At the US Navy Wave Energy Test Site at Marine Corps Base on the island of Oahu, Hawaii, several wave energy designs are undergoing testing. Azura, weighing 45 tons, looks like a giant steel ladder that floats vertically in the water, swaying back and forth like an inverted pendulum. A float in the middle bobs up and down in the waves driving a hydraulic motor that generates electricity.

The Lifesaver, a ring-shaped device, is tethered to the sea bottom by cables. As it rocks on the surface, winches attached to the cables are spun back and forth driving generators. Other open-ocean floating devices use the up and down motion of passing waves to pump air or water in and out of chambers, all trying to turn the motion of the wave into pneumatic pressure that can be run through a turbine generator.

Wave energy has tremendous potential, not just because there's so much of it flowing across the oceans but also because waves roll day and night, giving them an advantage over solar power. However, it is not totally reliable since wave height can vary from day to day and season to season, and with the passage of storms.

Devices designed to capture all that energy potential in waves are mostly that: design concepts. The challenge of making them commercially viable is huge because of difficulties working in a marine environment, maintenance costs, and storm risks.

If you want to do things on the ocean, there are challenges and opportunities. Firstly, the ocean likes to eat things. So corrosion is a significant issue. Survivability is a significant issue. Secondly, costs. Whenever you want to deploy, recover, or do anything to your device, you need to hire an expensive vessel, take it out there, anchor or moor it, and you have to hope that a big storm doesn't come along and blow it away. It's a lot more expensive and difficult currently.
Dr. Bryson Robertson

Unlike other technology, such as ships, which are designed to cut through waves and dissipate their energy, wave generators try to absorb all that force. That means they must be tough, able to resist corrosion and withstand storms at sea, barnacle growth, and collisions with floating debris. So far, prototypes are showing that it is technically possible to capture wave energy, but experts say the technology today is where wind and solar were in the 1980s, with at least a decade more research and funding needed to become viable.

So how will wave energy fit in with our green future?

Despite serious challenges, waves do have their advantages and the potential to take a place in our energy portfolio.

We need diversity as we move forward with global decarbonization. We need a little bit of everything, and wave has a unique signature of when it produces, how it produces, and where it produces electricity.

One, it produces when solar doesn't, because the sun doesn't always shine. Especially during winter in the northwest, the chances for solar are very low. And that's when we need the power.

Wind goes up and down like nobody's business. Wave, we can predict it really well, you can design the whole electrical system around renewables that are varying. Wave varies on a day scale, not on an hour scale like wind does. So that is a huge advantage.

How long do we want to develop these megaprojects way out on the flat lands then run these expensive transmission lines back to all our coastal resources? Because that's where people are living.

But if you can generate power off the coast, close to where your demands are, you're not wasting power by running it through long cables.

Dr. Bryson Robertson

Despite most of our planet being covered by oceans, wave energy is not likely to develop into megaproject size anytime soon. It will appear at a smaller scale in specific locations with designs suited to that spot and serve a small area, contributing to the local grid. To succeed, projects must overcome all the technical challenges of dealing with the marine environment, the mechanical complexity of turning the limited motion of water in a wave into a spinning electric generator, and then the social-acceptance factor.

Floating wave generators will have to be clearly marked with bright paint and warning beacons to avoid collisions with ships. Negotiations with local fishers will have to settle on the best locations to avoid entanglement in their nets. Then there is the visual element. Will local people living along coastlines want to see a flotilla of brightly coloured devices interrupting their view of the sea?

Waves are mesmerizing to watch, endlessly caressing sandy beaches or crashing violently against rocky headlands. They can be playful for surfers, or powerful enough to swallow large ships. Waves are the manifestation of pure energy in motion, an energy untapped and, for now, an element of nature to be mostly admired.

Reversing Falls, Saint John, NB.

Chapter Five

Tidal Power

Putting the moon to work. Update on the Bay of Fundy and other projects.

On Canada's East Coast, the Bay of Fundy is famous for having the highest tides in the world at 15 metres, the height of a five-storey building. This dramatic effect is due to a unique resonance between the shape of the huge bay and the rhythm of the tides. The ebb and flow of 160 billion tons of sea water, more than the flow of all the rivers in the world, is like water sloshing in a bathtub. Water funnelled into the bay causes a buildup at high tide that floods the shores, then at low tide recedes leaving behind vast mud flats.

This phenomenon became real during a sailing trip to the region when I pulled into a local harbour for the night. Soon after dropping anchor in what I thought was a safe spot, a local resident drove out to the end of the town wharf and shouted across: "You don't want to anchor there. That spot is going to be high and dry in two hours. You want to drop anchor over there," pointing to a spot less than 100 metres away. With great respect for local knowledge, I moved the boat to the appropriate spot and, sure enough, looked back later to see red mud where I had been floating a short time earlier. Tides are taken very seriously in the Fundy region.

Extreme tides can cause another phenomenon: where the high tide levels are above the level of a river that runs into the sea, the tide can temporarily reverse the flow of the river until the tide drops and the flow returns outwards. The most famous and most powerful is the Reversing Falls in Saint John, New Brunswick, a torrent that runs in both directions, where you can see the true might of the tides. This also became a terrifying place when I passed through . . . in a sailboat.

Many first-time visitors to the Reversing Falls are disappointed because it is not a waterfall at all, but rather a narrow gorge near the mouth of the St. John River where the water changes direction twice a day. During low tide, the river becomes a whitewater rapids with powerful whirlpools and standing waves. As the tide rises, the flow slows, then the river becomes completely calm for about twenty minutes until it reverses direction and becomes an equally raging rapids flowing upstream in the opposite direction. It is only during those brief periods of calm during slack tide that vessels transiting up and down the river can pass through, and you can't be late. Once the tide turns, the only boats that can handle that water are high-powered jet boats that used to give tourists thrill rides during the summer or experienced whitewater kayakers.

After sailing into Saint John Harbour, I caught up with some local sailors who were waiting to make the passage through the falls to a marina on the upstream side. They convinced me it was a piece of cake and to simply follow along, which provided the confidence I needed to make it through without incident. Motoring through the passage was a strange experience, seeing the river from the water perspective. I lived in Saint John for a while as a kid and used to visit the site at both high and low tide so I knew how rough it could be.

The following day, the return trip was a different story.

The section of river that makes up the Reversing Falls is less than a kilometre long. The water upstream and downstream of it is completely calm and navigable anytime. The journey from the marina to the falls was a few kilometres, so I had to time my departure to arrive at the falls at the exact right time to catch slack tide. Unfortunately, a sailboat under auxiliary diesel power is not very fast, only about 6 knots, which is about the speed of a bicycle, and I underestimated how long it would take to get to the falls by about fifteen minutes. That left only five minutes to make it through before the current from the rising tide would become too strong for my boat's single propeller to push against. In that case I would be forced back upstream and have to wait six hours for the next slack tide.

Rounding the final bend in the river, the expanse of the killer water stretched out before me. The surface was flat, but eddy currents were beginning to form here and there, stirring up the water with little pools like the river was beginning to boil. There were no other boats in sight. Everyone else had the sense to arrive on time and had already made it through. At the far end was the narrowest part of the gorge, spanned by two side-by-side bridges, one railway and one automotive. All I had to do was get under those bridges and I would be safe. I decided to go for it. If the current became too strong, all I had to do was back off and let it carry me back upriver to the marina. After all, I had already been through in that direction the day before and knew that, although the water could be rough, it was deep.

Right at the narrowest part, directly under the bridges where the flow of water would be fastest, lay a boulder the size of a pickup truck protruding through the surface almost in defiance, blocking part of the passage. An experienced sailor at the marina told me to aim directly for the rock rather than try to go around it. He said the current would push the bow of my boat safely around, otherwise I would be smashed against the walls of the gorge.

As I approached the narrow passage, I could feel the current building, slowing the boat down. My clock showed that the twenty-minute window was already closed, but I was very close to making it through, only about 50 metres to go. Just ahead was the boulder and a tongue of water beginning to wrap around it, growing stronger every second. I could see the rising tide coming in, flowing down towards me with increasing menace. Against my sailing instincts, I aimed the bow directly at the rock. Just when it looked like I would be crunching into the rock, the boat was carried to the right by the swift-flowing water like an invisible hand guiding my keel. Ahead was an actual hill of fast-moving water that I had to climb. A few metres off to the left, in the wake of the big rock, huge whirlpools swirled, one of them twirling a big log like the baton of a majorette. That log was almost the length of my boat. I couldn't get caught in that deadly vortex.

By this time my engine was straining at full throttle against the increasing current. I was under the bridges and had to watch the steel structures to confirm I was still creeping forward at an agonizingly slow pace. The safety of flat water was just ahead. All I had to do was hold the course between whirlpools and hope my engine was strong enough to make it through.

After what seemed an eternity, with engine screaming, the bow finally crawled over the crest of the tongue of water and into the safety of the lower river. Looking back, the entire passage I had gone through was completely white water. I looked up to the rim of the gorge where a crowd of tourists was looking down from an observation platform, cheering my success. I waved back, thankful for having made it through, with a greater appreciation for the power of the tides and a resolve not to use my sailboat as a white water raft ever again.

In the days of sail, when ships were true sailing vessels without the aid of modern auxiliary diesel engines, mariners relied on the ebb and flow of the tides to bring their wooden ships in and out of harbours. The phrase "go with the flow" comes from the tradition of riding the currents that flow out of harbours as the tide recedes, currents strong enough to flush even the largest ships right out to sea, or not. Sea captains are always aware of the tides, which is why departure times were strict. The big heavy vessels were often towed off the wharves and through the harbour by men rowing longboats, so they needed all the help they could get. If the tide was flowing in the wrong direction, the ship was going nowhere.

To anyone who has been to a seashore, the tides are something that seem to come and go throughout the day. One moment the waves

are rolling on a wide sand beach, then the next time you look the beach is gone and the waves are crashing against rocks or a breakwall. Unless you sit and stare at the sea for hours, you don't usually see the change happening. And in the daily movement of that water there is energy. On a global scale, there is an estimated 100 gigawatts of tidal energy available, enough to power eighty million homes. But that energy is very site specific, such as the Bay of Fundy on Canada's East Coast where the tides are the highest in the world, or where ocean water is funnelled among coastal islands producing rapid flows. Ocean water is eight hundred times denser than air so it carries far more energy per unit volume. That means underwater turbines can be much smaller than wind counterparts to capture the same amount of energy.

Tidal energy has a huge advantage over other renewables: it is entirely predictable down to the minute for thousands of years into the future thanks to the clock-like precision of the movements of the Earth, moon, and sun.

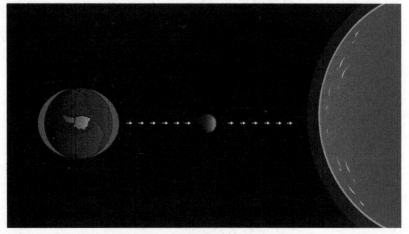

Tidal bulge caused by the gravitational pull of the moon and sun.

Image courtesy of SciJinks, NOAA

Tidal energy is driven by the gravitational pull of the moon and the sun, which distort the shape of the oceans and even the body of the Earth itself. Twice a day, the ground beneath your feet rises and lowers about a metre because of Earth tides. This motion is so smooth and spread over such a large area that we don't feel it. A greater effect is felt by the oceans, which are more fluid and move more easily.

The moon literally lifts the ocean in a lens-shaped bulge that forms on both sides of the Earth, the one facing towards the moon and the one facing away. These bulges can vary in size depending on the position of the moon in relation to the sun. When the moon, Earth, and sun line up, their combined gravitational pull raises the bulge higher than when the moon and sun are out of alignment with each other.

These tidal bulges are always there, and the Earth turns under them. If the Earth didn't spin on its axis every twenty-four hours, there would be no tides. Think of it as the gravity of the moon holding the ocean up as the body of the planet turns beneath it, like walking within a tent being held up by a pole in the centre. When you are under the peak and the roof is highest over your head, that is high tide; as you move to the edge of the tent, the roof is lower and that is low tide. Rather than thinking of tides as water moving up and down, think of it as the ground moving under a bulge of water that is held up by the moon.

Tidal flows vary greatly around the globe, depending on the geography of shorelines. Island countries in the middle of the deep ocean do not experience very high tides because the surface of the ocean doesn't move up and down more than a metre or so. But along coastlines with deep bays or around a cluster of islands, the tidal flow can become amplified and flow very strongly when it funnels through narrow channels or piles up in basins.

All this movement of ocean water is reliable, predictable, carbon-free energy that can be captured and turned into electricity. But it only works in regions with the right geography to create these strong flows. Interestingly, the Reversing Falls is not one of those locations. Perhaps the flow there is too strong for tidal technology to handle. But the Bay of Fundy is still one of the prime locations to capture the energy of the tides.

According to Dr. Curran Crawford, at the Institute for Integrated Energy Systems at the University of Victoria, who worked on tidal projects, the potential of tidal power is huge, but until recently, the challenge to capture that energy has also been huge.

Water is a really dense fluid so you can get more compact systems for a given power output, but with tidal, the total, actual harvestable amount is not that great. You end up with very specific sites where you can use it. Nova Scotia and BC have a number of sites.

As soon as you take technology offshore, the costs are quite challenging. I think it was oversold. People tried to go megawatt scale with both wave and tidal, but they had to scale back expectations and operate in the 100- to 500-kilowatt scale. You can have a good contribution at specific sites at that scale in particular, remote communities, but it's not a panacea.

There are two approaches to capturing tidal power. One is to build a barrage, which is basically a dam that takes advantage of the change in height between high and low tide to fill the area behind as the tide rises, then allow it to flow out through turbines as the tide falls. The second method is to use turbines on their own that act as underwater windmills, mounted either directly on the sea floor or as propeller-type devices hanging off floating barges that are anchored in the tidal flow.

Tidal Barrage

The tidal dam, or barrage, is basically a barrier in the mouth of an estuary that refills itself twice a day. Sluice gates are opened during high tide, allowing water to flow into the catch basin upstream. That basin is also fed by a river. The gates are then closed, and water is held back until the tide drops the water level on the downstream side. When the difference in water level between the two sides of the barrage reaches 7 metres, gates are opened and the water flows through a tunnel containing a turbine, which is a large, four-bladed propeller that spins to produce electricity. This means the system is only running during low tides, so there are gaps lasting several hours twice a day when the electricity is not flowing. More modern systems make up for some of that gap by using reversible turbines to catch the flow of water running in both directions through the barrage.

However, the cyclic nature of tidal energy means there needs to be some kind of backup power source to keep the lights on during slack tide. On the positive side, these down periods are predictable right down to the minute, unlike with wind or solar, which can be interrupted any time by changes in the weather.

The first large-scale tidal barrage was the Rance Tidal Power Station built in France in 1966 with twenty-four turbines and a capacity of 240 megawatts. It was the largest in the world for forty-five years and is still in operation today.[1] It has now been superseded by the Lake Sihwa project in South Korea, which provides power twice a day at high tide. Sluice gates in a long seawall are closed at low tide. Then when the sea water is at maximum height on the outside of the wall, it is allowed to flow into the lake through ten turbines, each with a capacity of 25.4 megawatts. The plant supplies enough power annually to support the domestic needs of a city of five hundred thousand.

The only tidal dam project in North America, and the third largest in the world, was at the mouth of the Annapolis River in Annapolis Royal, Nova Scotia. This project was developed in response to the energy crisis of the 1970s and began operating in 1984, with a capacity of 20 megawatts, delivering 80 to 100 megawatt hours of electricity a day depending on the tides. Built as part of a causeway, the dam delivered all that power using a single horizontal turbine measuring 7.4 metres in diameter.

The Annapolis project operated successfully until 2019, when maintenance issues shut it down, which became somewhat of an excuse to cancel the project for environmental reasons. The difficulty with these types of barrages is their effect on marine life and the surrounding ecosystem. Studies found that the presence of the dam caused erosion upstream, but the most visible impact came when two humpback whales managed to make their way through the gates at high tide and became trapped behind the dam.

In 2004 a mature humpback was trapped for several days, earning the nickname Sluice. Fortunately, it was able to find its way back out through open gates. But a few years later, the corpse of another juvenile whale was found in the river upstream. It is believed that the whale followed fish through the gates but could not find its way out. A barrage-type tidal power project would likely have difficulty passing environmental approval today.

Underwater tidal turbine.   Image courtesy of Simec Atlantic Energy

Turbines in the Flow

A less invasive and less costly way to capture tidal energy is to put turbines directly into the tidal flow without the use of dams. In areas where tidal currents are funnelled between islands or channels, the flow is accelerated and can be tapped by the equivalent of underwater windmills. They even look similar with two- or three-bladed designs.

Experiments in Scotland, Nova Scotia, and British Columbia are taking two approaches: one, where turbines are placed on the sea bottom and, the other, where they hang below floating barges. Bottom-mounted turbines placed among the Orkney Islands of Scotland used blades spanning 14 metres capable of generating 2 megawatts of power. These prototypes operated for several years in the mid-2000s to evaluate performance and, more importantly, their impact on marine wildlife. Of course, any moving mechanical structure placed underwater will suffer from the ravages of corrosive salt water and growth of marine organisms. Special slippery paints were tested to keep the rotors free and clear.

Since then, four turbines off the north coast of Scotland generated enough energy to power nearly four thousand homes in 2019. MeyGen, the world's largest tidal array, is the first phase of a project that could eventually power 175,000 homes with more than 250 sub-merged turbines.

Extensive monitoring of sea life, especially mammals such as seals and whales, must be tracked to determine whether the presence of the turbines affects their behaviour.

In early studies, researchers found that the animals tended to avoid the turbines because of the low-frequency noise generated, and no deaths were recorded. On the other hand, any underwater noise produced by technology can interfere with the sounds whales them-selves produce for echolocation to find food. More research is needed to ensure that future installations do not interfere with whale migra-tions or sea mammal feeding and breeding areas.

Fortunately, from an environmental perspective, many areas where tidal currents are the strongest are also unsuitable for marine life. The powerful currents tend to scour the bottom so there isn't a lot of food, and fish and other creatures don't congregate as much in those areas. But at the same time, those conditions present a chal-lenge for tidal project technology, both from an engineering perspec-tive and economically. According to Dr. Curran Crawford, that is not always easy.

> I thought underwater would be nice smooth laminar flow, but it's not. It's nasty eddies and turbulence in there that can snap your blades off. So we can design for that now that we know what the loads are. But it is really just trying to get the costs down. It all comes down to what your costs are per kilowatt hour delivered.[2]

A less costly method of capturing the energy of the tides is the floating barge. Actually, "barge" is the wrong term because these structures look more like ships or submarines that are permanently anchored. Again in Scotland, one built by Orbital Marine Power looks like something Batman would design. With a capacity of 2 megawatts, it looks like a giant yellow submarine 72 metres long floating on the surface of the water with large wings on either side with huge pro-pellers 20 metres in diameter on each wingtip. The device can be towed to any location, where it is anchored to the bottom using four

Orbital Marine tidal generator looks like something Batman would drive.

Photo courtesy of Orbital Marine Power

heavy-duty chains. The wings are lowered into the water and the turbines begin to spin. The blades of the propellers are reversible so they can catch the tide running in both directions. This project has been in development for fifteen years, and the company hopes to make these generators available anywhere in the world.

A similar concept is being tested in Nova Scotia, but instead of a single hull, the device uses a slender trimaran with six turbines hanging down from the stern. From below, they look like propellers to drive the boat; however, the boat remains stationary tethered to the bottom, while the tides spin the blades.

The advantage to the floating turbines is ease of maintenance. Large structures resting on the bottom must be raised by crane for maintenance, which is very expensive, but the floating barges can lift their turbines out of the water for service work. The trimaran design has the blades mounted on lever arms that can be rotated up, like outboard motors, so they stick out above the water right off the stern for easy maintenance. They also automatically rotate up if one of them is struck by a large object such as a floating dead tree.

Parallelling the development of tidal energy projects are sophisticated sonar systems to track the movement of whales, especially

Sustainable Marine Energy's floating tidal turbine being tested in the Bay of Fundy.

Photo courtesy of Sustainable Marine Energy

endangered right whales that populate the Bay of Fundy. Care must be taken to shut down the turbines when whales are approaching.

The episodic nature of tidal power, where the water stops flowing four times a day while it changes direction, requires the use of energy storage systems or backup power to keep the flow of electricity constant. While many experimental prototypes of tidal turbines have been tested, none have gone to full-scale production the way wind and solar have.

Overcoming Hurdles

According to Dr. Lekelia (Kiki) Jenkins, associate professor in the School for the Future of Innovation in Society at Arizona State University, who studied tidal power in Puget Sound in the Pacific Northwest, the obstacles to tidal power in the United States are as much political as they are technical.

> The greatest obstacle is a thing called the Valley of Death, which is the stage when you have your initial prototype, you get it in the water, but then you have to scale it up and get your energy services to the grid.

We see that wave and tidal projects die at the point of getting an array of larger prototypes in the water and contributing to the grid. Why is that happening?

It is happening partly because wave and tidal are being developed by private companies. They need a return on investment, and there is a disconnect between where they can do this scaling up in terms of how we govern our ocean resources, the laws and regulations that are in place. There is not a space for them to develop from prototype to large commercial developments without certainty that they will be supported by governments.

The technology isn't proven well enough for them to take this amount of financial risk. Our regulatory system is decades behind where the technology has gone. If we can't regulate the technology, the private funders are not going to support it if there is no place in the water to put them.[3]

While federal regulations may be an obstacle in the United States, that is not the case in Scotland, where tidal power is receiving a lot of support from government and industry.

Scotland is interesting because there is a strong independence movement alive there. Their marine renewable resources are phenomenal, in a system where there are not a lot of other prospects for diversity. They see this as a way to become economically independent from England.

To have their own power base—to have enough to sell it to the rest of Europe, and have the independence, have the jobs— is an incentive to have it come from tidal. So it is very compelling for them to figure it out, for the government to create legislative structures and provide large innovation spaces where they can do the work. We don't have that in the US.
Dr. Lekelia Jenkins

Tidal power may be best suited on a smaller scale to specific remote locations where local power is needed. In Haida Gwaii, a group of islands off the coast of British Columbia, a novel system incorporates energy storage so no backup energy supply is needed.

Haida Gwaii, like many remote communities, relies on diesel generators for power. Millions of litres of fuel are burned each year, contributing to climate change. A new tidal power concept involves a four-bladed paddlewheel whose only purpose is to pump water. When the tidal flow turns the wheel, water is pumped uphill to a small storage tank, which then allows the water to flow back downhill through pipes to a turbine that makes electricity. In addition, the capacity of the tidal pump is greater than the electrical turbine needs, so some of the water continues farther uphill to a larger reservoir that is filled during tidal flow, then drained down to the turbine during slack tide when the pump stops. The engineers hope this design could free the community of fossil fuels entirely.

Underwater turbine technology has also been proposed for large rivers such as the St. Lawrence where the flow is more constant. But again, studies must be undertaken to ensure that these underwater propellers don't become chopping machines for fish and whales or hazards to navigation for shipping.

Finally, at least in North America, there is the issue of public acceptance:

> On one hand, people say this is renewable energy and that's good for the climate. On the other hand, they are concerned about salmon and orcas and are uncertain about the impact of these technologies on the environment and don't want to mess with it.
>
> There are questions about putting technology out onto the oceans because they have hydroelectric, which you can consider green and renewable. People with the environmental ethos are actually on both sides, for and against tidal. It depends on what they are favouring, whether it is ecosystems and animals or climate issues. It also comes down to what they know and what they understand, because this technology is not something that is talked about in the media. They don't see it around them. There are analogies about chopping fish—which are actually inaccurate—but if you google it, that's what you come up with.
> *Dr. Lekelia Jenkins*

As more coastal towns become tourist destinations, people want to look out and see the deep-blue sea, not an array of tidal generators. Submerged turbines have the advantage of being out of sight, although their location must be clearly marked to avoid ship collisions and entanglement in fishing gear. While it is a more expensive way to generate electricity than wind and solar, the technology is evolving, so tidal power will have a place in specific locations, especially remote areas and island communities.

What is most remarkable about tidal power is that it is the only source of energy on Earth that comes from the moon—and the moon will circle the Earth for billions of years to come. People living in parts of the world where tidal power is operating will be able to look up at that romantic glowing orb in the night sky and marvel at the fact that it is also providing the energy for the glow of the lights in their homes.

Myth of Sisyphus.

Image courtesy of Pinclipart

Chapter Six

Energy Storage

Flow batteries, reversible dams, underwater pressure—
technologies needed to store energy when the wind
doesn't blow and the sun doesn't shine.

Leave it to the Ancient Greek gods to come up with clever ways to punish those who go against them. Sisyphus, one of the tragic figures of Greek mythology, was a former king of Ephyra, now Corinth, who was forever condemned to roll a large rock up a hill only to have it roll back down once he reached the top. It was a futile task he was doomed to repeat for eternity, all for the sin of trying to cheat death and offending the gods. Philosopher and author Albert Camus, in his essay "The Myth of Sisyphus," felt the former king could take some joy from the fact that as the rock rolled back down the hill, he was temporarily free from that burden and could transcend the meaninglessness of his life as he walked down the hill free from struggle. As he says in his essay: "One must imagine Sisyphus happy."

This is the ultimate case of when life hands you lemons, make lemonade.

Although he probably didn't appreciate it at the time—and for that matter, probably still doesn't—there is one more reason Sisyphus could take comfort in the efforts of his endless task. He was the world's first battery.

That's right, Sisyphus the rock-roller was a battery. We tend to think of batteries as the large, heavy black cubes that run our cars, the small tubular penlight batteries that run our remote controls, or the myriad lithium ion batteries that run our devices. All of them store energy in chemicals that can be turned into electricity. But other devices can behave like batteries by storing different forms of energy, such as mechanical energy in the coiled spring of a windup toy, hydraulic energy in water held behind a dam, heat energy in a hot water tank, or potential energy in a heavy object raised to some height. All these energy storage systems can be converted into electricity and put to useful work.

The act of rolling the large boulder up the hill requires a lot of muscle power from Sisyphus, which is put into the rock in the form of potential energy as he pushes it higher and higher. If the boulder remained at the top of the hill, which would please Sisyphus immensely, it would store all that muscle-power-generated potential energy, the way a fully charged battery sits quietly until we need electricity to start the engine in a car. Water raised behind a dam sits peacefully as a lake until the penstocks are opened and it rushes through turbines to generate electricity. Unfortunately for Sisyphus, his big boulder does not sit quietly at the top of the hill for long: it discharges itself by turning potential energy into kinetic energy—the energy of motion—and rolls down the hill. By the time it stops at the bottom, all the energy has been released and the system is back to zero, like a dead battery. When the unfortunate Sisyphus starts his long, arduous push upwards again, he is putting energy back into the rock. His rock battery is rechargeable!

Energy storage is a vital and often forgotten element of our move to an alternate energy future. We have been spoiled by our current energy generators: whether driven by fossil fuels, nuclear power, or hydroelectricity, they can all run 24/7, so energy is always there when we need it. But devices like wind turbines and solar panels operate at the whim of the weather, where the wind doesn't always blow and the sun doesn't always shine, so their energy output is variable or sometimes non-existent.

To compensate for this major drawback, on calm and cloudy days when the power is not flowing, solar and wind farms are often accompanied by so-called peaker plants, small fossil-fuelled generating stations, usually powered by natural gas, that can be turned on quickly to fill in those energy gaps. So then the argument goes: If these wind and solar farms need fossil-fuel backups, they're not really clean. Why not just go with natural gas alone?

The difficulty is that electricity itself is not easy to store. It won't just lie around in waiting; it needs to be used as soon as it is made, like the flame in a campfire that needs to be constantly fed by wood. To store electricity, it needs to be converted into some other form of potential energy, then converted back into electricity when needed.

When you think about it, the gas-fired peaker plant is really an energy storage device. The natural gas stores chemical energy that is released when the plant is fired up. But natural gas is a fossil fuel that produces climate-warming carbon dioxide. The solution is, rather than use fossil fuel as a backup, use another type of energy storage device that can be drawn upon when needed. One way is to tap some of the energy generated when the wind is blowing hard or when electrical demand is low and stash it away in some form that can be used to fill in the gaps. Fortunately, energy can be stored in many different forms, some of them surprising.

The electrical grid, the huge network of wires that run from generating stations across the land, through cities, and into the walls of our buildings, is like a living organism, a pulsating nervous system that must be kept alive, humming at the same frequency everywhere at once and balanced throughout. Generating stations pump life into the system while our homes, workplaces, and industry take energy

City lights outline the electrical grid, a web of energy that spans the continent.

Image courtesy of NASA

out. Once the electricity is produced and is flowing in wires, it has to be used immediately, which means the lights above you at this moment are powered by electricity that was generated moments ago in a facility not far away. It is also amazing that the amount of electricity reaching your home is just right so your devices can handle it. Too much is a problem, as you find out when circuits become overloaded and either start fires or shut down and cause power failures. Too little, and you get more blackouts. It is a marvel of engineering that the whole network can be kept alive all at once to meet everyone's demand. To keep it running smoothly, it is all about balancing the energy flow in and out. But what if there's more in than out?

There are two solutions to the problem of what to do with excess power: one is to move it somewhere else, or convert and store, then move it later.

In the past, excess electricity coming out of generating stations was sent down long-distance transmission lines, losing energy from electrical resistance and creating waste heat—another form of throwing energy away. Today, power can be shuttled within the grid itself to areas that need it through an intelligent control system called the smart grid.

Intelligent grid management systems monitor energy supply and demand on a fine scale, tracking where energy is needed and where it is in excess. By controlling the flow through the grid, the balance of power can be equalized. Together with other forms of energy storage, they can flatten out that duck curve and provide reliable continuous power day and night. In the future, smart grids will be managing energy flow from many different sources, from large-scale energy farms to the solar panels on the roofs of houses.

If we are to truly wean ourselves from fossil fuels and depend more on clean alternatives, this problem of variable power output, where electricity is plentiful during certain hours and lacking in others, will become greater. The second solution is to store energy when it is plentiful and use it later when needed.

Although electricity itself can't be stored easily, it can be converted to some other form of energy—chemical, pressure, or gravity-dependent—that will sit still for long periods of time.

We've been storing energy throughout civilization. Wood stores chemical energy so when a little heat and oxygen is added, we get a crackling fire. It has been cooking our food and heating our dwellings for millennia. Then we discovered other energy storage systems that come out of the ground, the fossilized remains of plants and the animals that ate them millions of years ago. First it was coal—literally rocks that burn. These black rocks burned hotter than wood and lasted longer, so you didn't need as much to get the same amount of energy. In other words, it was more energy dense. Then came oil, which was cleaner burning and more convenient than coal because it could be easily stored in tanks and carried around in vehicles. Along with natural gas, these fossil fuels, storing huge amounts of chemical energy, powered the Industrial Revolution.

Today we are witnessing another industrial revolution with the quest to find not only alternative sources of energy but also energy-dense ways to store them.

The Chemical Solution

The frontline energy storage system is batteries, which convert electrical energy into chemical energy and back again. They are the most convenient because they can respond instantly to changes in energy demand. Battery technology has not really changed much for more than two hundred years, when Italian scientist Luigi Galvani stuck two different metals into the leg of a dead frog, connected them, and made the leg move. He thought the electricity was coming from the frog itself, but it was his fellow countryman Alessandro Volta who believed it was the two different metals generating the current with the fluids in the frog acting as an electrolyte between them. To prove the concept, he built stacks of copper and zinc plates with paper soaked in salt water between them and found that an electric current flowed through a wire attached to both ends of the stack. It was the first battery. The unit of electric charge, the volt, is named after him. Schoolchildren do this same experiment today in classrooms using a galvanized zinc nail and copper penny stuck into a lemon. The acidic lemon juice acts as an electrolyte, and when a wire with a small light is connected between the two metals, the light illuminates.

How a Battery Works

A battery is simply two different materials, usually metals—such as lead and lead oxide in a car battery or lithium oxide and cobalt in a lithium ion—with an electrically insulating liquid, or electrolyte, between them. Chemical reactions between the electrolyte and the metals remove electrons from one side and add electrons to the other. The side that gains electrons, the anode, develops a negative charge; the side that loses electrons, the cathode, a positive charge. During this process, both metals and the electrolyte change chemically. The cathode is oxidized, which is basically rust, like the red coating that appears on iron, and gives up electrons. The anode accepts those electrons through another reaction with the electrolyte called reduction. The reaction in the electrolyte can produce hydrogen gas. The electrons want to flow back to the positive side and rejoin their atoms, but they are blocked by a separator, so they have to go around the outside of the battery, which happens when a wire connects the two sides. As electrons flow from the negative to the positive side—presto: your flashlight shines or your car engine starts. When the metals deteriorate and electrolyte is depleted, the chemical reactions stop and the battery goes flat.

Disposable batteries only do this reaction once, but rechargeable batteries pump electrons back into the system, reversing the reactions so the electrons move back from the positive to the negative side and the cycle can be repeated. But even rechargeable batteries degrade over time. The capacity of the battery depends on how many electrons can be stored in the anode, and how many ions the cathode can absorb. Intense research is underway to find the best materials for maximum capacity, quick recharge times, and least degradation—all at the lowest cost. Bringing all those elements together into a super-battery is the ultimate challenge.

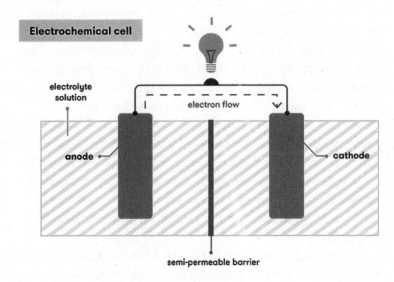

Cross-section of a battery showing the electron flow.

Image courtesy of the Australian Academy of Science

In the past, batteries have been made of cheap materials that hold enough charge to get the job done. The most common example is the large, heavy, brick-shaped lead-acid batteries, seen in conventional cars and trucks, that have been around for more than a century. They use lead and lead oxide as the two different metals with sulphuric acid as the electrolyte between them.

It is amazing to think that we have put up with these containers of toxic metals and poisonous acid to start and run our vehicles for so long. But lead is plentiful and cheap, plus in vehicles, the electrical demand on them is quite low; once the engine starts, it keeps the batteries constantly charged back up through the alternator or generator. Although anyone who has tried to start an engine on a sub-zero winter morning knows that the charge is often not enough to get the job done.

Attempts have been made to give lead-acid batteries more juice by ganging them together to store more charge. For example, on pleasure boats, one battery is dedicated to starting the engine while another, or several others, are used to run the lights, water pumps, refrigerators, and electrical outlets when the boat is not plugged in to shore power. But even these so-called deep-cycle batteries need to be recharged daily. Early attempts to use lead-acid batteries for electric vehicles ended up with a cluster of batteries that took up a lot of space, weighed more than the vehicle, and only provided limited range. But that is changing rapidly as electric vehicles become more popular and batteries are seen as one way to store electricity on a large scale to fill in when alternative sources of energy such as wind and solar are not producing. A new industry has emerged as companies around the globe are pouring billions into developing a wide variety of new materials that can store more charge. The challenge is to come up with a product that is energy dense, lightweight, and easy to manufacture at low cost. That is not always easy.

For decades, small flashlight-type batteries have been made of zinc and carbon with a carbon-manganese oxide powder acting as electrolyte. The alkaline battery, made famous by the Energizer bunny for its extra-long lifespan, uses zinc and manganese for the electrodes with an alkaline potassium hydroxide electrolyte that gives it its name. That technology was invented by Canadian engineer Lewis Urry in the 1950s.

Television news crews were given new freedom of mobility with nickel-cadmium batteries that made cameras portable so they could be on location for developing stories. These solid black bricks the size of your hand also eliminated electrical cords from power tools. But as convenient as they were, the now-ubiquitous lithium-ion battery overshadowed them. It has the highest power density of all previous batteries, which is why it drives just about every portable device you can think of, plus the electric vehicle industry gave it the boost needed to enter the market on a large scale.

Lithium is one of the lightest metals, and very reactive, which means it easily gives up its electrons to other materials, and that makes it ideal for batteries. It also provides a higher energy density, so a cell produces 4 volts compared to the typical 1.5 volts of an alkaline cell. There is no one form of lithium-ion battery. The lithium is combined with other elements such as cobalt or aluminum and used in both the anode and the electrolyte. Large numbers of lithium ions can be moved back and forth between the different sides of the battery during charging and discharging. Research is focused on finding the best chemistry to provide high energy density for long range, maximum number of cyclings between charge and discharge, and quick charging times.

Lithium ion has become the mainstay in batteries, but early versions were not without disadvantages. Early liquid electrolytes were flammable, and leakage resulted in fires both in phones and cars. (That's why airlines ask if you have any lithium-ion devices in your luggage.) These batteries can take a long time to charge, experience decomposition at higher voltages, and lose a bit of their capacity with every charging cycle, which can amount to one-third loss of the battery's capacity within a decade. There are also problems with dendrites that form in liquid electrolytes, crystals that grow like the stalactites hanging from the ceilings of caves. They can reach across between the anode and cathode to short the battery out, or even cause it to explode. Researchers have worked steadily to find ways to prevent those growths and extend the lifetime of the battery. This involves compounds such as lithium cobalt aluminum oxide, or NCAs, used in the Tesla Model S.

Recent versions in electric vehicles are approaching the energy equivalent of a tank of gasoline with ranges exceeding 300 kilometres and quick charging times of fifteen to twenty minutes.

Much of the research was pushed by Tesla, Inc., which showed that electric vehicles not only can be practical and attractive but also, with their incredible torque from the electric motors—torque that is there right from zero—and no need to switch gears, can outperform even the hottest gasoline-powered muscle cars at the drag strip. Now all the major automotive manufacturers are producing electric vehicles, with the most dramatic change happening at Ford. Its classic Mustang, the hot rod muscle car dating back to the '60s, was converted into an all-electric crossover, which *Car and Driver* magazine named the electric car of the year in 2021. It is converting their most popular pickup, the F-150, to fully electric as well.

Tesla has moved beyond electric vehicles to large-scale lithium-ion battery packs that cover hectares of land and provide 250 megawatts of instantly available power as backup for solar and wind farms. One of Tesla's largest battery packs in the world, in Australia, can power a small town for four hours on battery power alone.

Interestingly, the company has found that rather than build single big batteries, it is more efficient to combine many small ones, thousands of cells that look like the AA penlight batteries used in flashlights and portable devices. If problems occur, individual cells rather than the entire unit are replaced. This approach has worked in the battery packs for Tesla electric vehicles and has now been scaled up to Powerwalls for home use and to large-scale megaprojects.

A recent innovation is the Tesla 4680 lithium battery, which rolls the elements into a spiral shape inside the battery to increase surface area and ease the flow of electrons. It offers faster charging and greater energy density at lower cost. These new batteries are being produced on a large scale; the biggest factory, Giga Berlin in Germany, will be turning out over 100 gigawatt hours of battery energy per year.[1] That's huge.

Another approach to battery technology is to eliminate the liquid electrolyte altogether. The idea of a solid-state battery is to still use lithium, but the liquid electrolyte is replaced with a solid material and there is no second metal as an anode. The battery is flat and lightweight. When charged, lithium ions flow through the solid material and form a layer of pure lithium metal that grows on top with a negative charge. Unfortunately, this makes the battery expand and contract as it charges and discharges. This form of battery provides greater energy density that could double the range of an electric car, or make it go the same distance with a battery pack that is lighter and charges more quickly.

The energy density of batteries is measured in watt hours per kilogram (Wh/kg). Tesla's new 4680 battery has over 300 Wh/kg while the new solid-state batteries are expected to carry up to 500 Wh/kg.

The ultimate goal is to produce a "million-mile" battery. It would be nice if that meant it would go that far on a single charge! But the term refers to a battery that could outlast the lifetime of the car and be reused in a new vehicle, reducing the amount of battery disposal.

However, all new technology has its drawbacks. Some of these solid electrolytes are not as electrically conductive as liquids so there is a lot of research into developing novel materials, including glass. Having said that, major automotive companies are investing in this technology because it holds the promise of dealing with the major issues that have been holding consumers back, namely range, long charging times, and better performance at lower temperatures. Toyota plans to double the range of its electric vehicles with a 0 to 100 percent charge in ten minutes. If a battery could actually go 1,600,000 kilometres (1,000,000 miles) over its lifetime, that is far beyond the lifetime of a car. Most vehicles start to wear out around 200,000 km which means the same battery could theoretically be recycled in eight new cars.

All this new technology is still in the research and development phase with companies experimenting with different chemical elements to lower costs and improve performance. The race for the million-mile battery is so intense it is hard to keep up. It is only when these new batteries can be manufactured on a large scale that they will be able to compete. Toyota is not expected to be in full production with a solid-state battery until the mid- to late 2020s, while Tesla is already underway with its design.

One way around the slow charging problem is to add another technology to the batteries: the supercapacitor. They are similar to batteries in that they can hold an electric charge, but there are no complex chemicals or electrolytes, just two conducting plates, one positive and one negative. A supercapacitor can absorb a full charge in a matter of seconds, which is great when recharging a vehicle. The problem is, they also release the charge all at once, essentially giving you a lightning bolt, which is not great when you want to go a long distance. One idea is to have a supercapacitor absorb the charge from regenerative braking, when the movement of a vehicle as it comes to a stop is used to pump charge back into the system. Then the supercapacitor can give back that energy as a quick boost to accelerate the vehicle when it starts up again.

The big question is cost. At the moment, as with any new technology, solid-state batteries and supercapacitors are more expensive, but if manufacturing can be streamlined, costs will come down and a new era of battery technology could begin rendering gasoline engines obsolete. The time may not be far off when gas-powered cars only reside in museums alongside steam engines.

Go with the Flow

A battery is like a bucket that can only hold so much water. If you want to store more energy, you come to a point where you need more batteries, which take up more space and add weight. One concept to get around that problem is the flow battery, which uses liquids to carry the charge. Think of a battery with a fuel tank attached to it. Rather than storing all the energy within the space of the battery itself, a flow battery keeps the charge in external tanks containing electrically charged liquids that flow through the battery with a separator between them so that they act as a liquid anode and cathode to produce electricity. As long as the liquids are flowing, the battery will operate, and the whole system is reversible. Although the energy density is lower than lithium-ion batteries, liquid tanks can be very large, so the storage capacity can be huge. This technique would obviously be best suited for stationary energy storage, such as a wind farm that has a tank farm in the middle. Some of the energy from the turbine would be pumped into the tanks to keep them fully charged so that the stored energy can be released when the wind stops. The amount of energy that can be stored is simply a function of the size of the tanks, which in an outdoor setting could be as large as you care to make them.

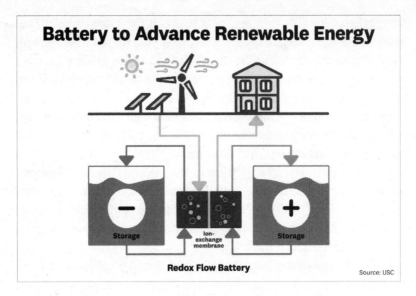

Battery to Advance Renewable Energy

Redox Flow Battery

Source: USC

Flow batteries store energy in electrically charged liquids in storage tanks to act as backup when the wind doesn't blow.

Photo courtesy of University of Southern California, Kandace Selnick

Flow batteries such as the redox have their issues, such as the need for a rare and expensive element called vanadium that is held in a toxic acid solution. Researchers are looking for alternative liquids including organic compounds such as quinones that can be derived from living materials such as wood. It is also much cheaper.

On a smaller scale, theoretically, a tank the size of a hot water heater could store enough energy to power an average household during times when energy prices are high or to supplement solar panels on the roof. While flow batteries are currently more expensive than lithium-ion, they have great potential if liquids can be found that are stable, cheap, and capable of many cycles of charge and discharge. This would make them cheaper than lithium-ion batteries over thirty to forty years.

Sisyphean Ideas

Getting back to Sisyphus and his rolling rock, gravity comes in handy when it comes to storing energy, especially since we spend so much time fighting it every time we want to go up the stairs, up a hill, up in an airplane, or up in a rocket. Like Sisyphus, we put a lot of work into moving things vertically. But as the old saying goes, what goes up must come down, and if we catch what is coming down, we can get back most of the energy we put into it.

When you place a heavy potted plant on a high shelf in your home, it takes energy from your muscles to lift the plant up, but once the pot reaches that shelf, what your muscles put into it is stored in the form of potential energy. That is, the potential to fall. If the pot remains on the shelf, the energy is stored, but if someone bumps against the shelf and knocks the pot off, the potential energy is released with enough force to shatter the pot, scatter the contents, and take a divot out of the floor.

Now imagine controlling the fall of the pot so its potential energy is released over a longer period of time. Perhaps the pot is attached by a string that is wound around a wheel so the wheel spins as the pot falls. If the wheel has a generator attached to it you could make electricity while the pot is falling, or if the wheel has fan blades on it, for a second or two you get air conditioning for your house. The challenge is to extend the fall of the pot for as long as possible, perhaps through a set of gears that slows the pot's fall so the electricity will flow longer. You could also let the pot fall from a greater height, say, the balcony of a thirty-storey building.

Controlling the fall of heavy weights to extract their energy is not a new idea. Grandfather clocks that stood tall with a long pendulum swinging inside producing the somewhat eerie tick-tock-tick-tock were powered by gravity. Winding up the clocks raises weights inside that are suspended from long chains. The chains wrap around a shaft, but rather than allowing the shaft to spin rapidly as the weights fall, a device called an escapement catches the teeth of a gear on the end of the shaft and only allows it to turn slowly, one cog at a time, step by step in one-second increments. That's what makes the tick-tock sound. The timing of the lever is controlled by the pendulum that swings back and forth with a one-second period. So, the pendulum controls the timing, and the weights turn the gears that move the hands of the clock. A little bit of the energy from the weights is also added to the pendulum so it keeps swinging and doesn't slow down

from friction. Grandfather clocks operate using gravity by controlling the fall of heavy weights.

Now scale that principle up about a million times to generate
electricity.

The Energy Vault tower uses gravity to store energy. Photo courtesy of Energy Vault

Two design concepts for mechanical batteries involve moving
large weights up and down a great height. One in Switzerland called
Energy Vault involves a 150-metre-tall tower, that's about thirty-five
storeys high, made of very heavy blocks stacked on top of one another.
Each block, made of recycled building materials, weighs 35 tonnes
and there are six thousand of them stacked into an interlocking
cylindrical shape. When energy is needed, three construction cranes
at the top of the tower pick up blocks and allow them to slowly fall
towards the ground. The electric motor that was used to haul the
blocks up becomes a generator powered by the falling blocks to generate electricity. Several blocks can be dropped in sequence to keep
the electrical flow constant and meet different demands. The company claims it could provide 35 megawatt hours or more of electrical
energy storage. The mechanical battery can respond to electrical
demand within milliseconds and can provide power for up to ten
hours, which makes it comparable to lithium-ion batteries. When all

the blocks reach the ground they are carefully stacked until electricity from a local wind farm is used to haul them back up again during off peak hours. A rechargeable mechanical battery.[2]

A variation on the theme is produced by a Scottish company, Gravitricity. Their system, of the same name, raises and lowers heavy weights through abandoned mine shafts. Abandoned shafts can be kilometres deep and are available all over the planet. The company is claiming a fifty-year lifespan with no degradation in energy storage, which makes sense, because a multi-tonne weight today will still be the same multi-tonne fifty years from now. The beauty of the underground system is that there will be very little to disturb the landscape above ground, unlike thirty-storey towers and cranes. The company claims it will be able to supply 25 megawatt hours of energy storage using a multiple weight system. That could provide emergency power for 10,000 homes for one hour.[3]

Sisyphus would be amazed by these gravity machines.

If dropping weights is not your cup of energy tea, how about spinning them?

If you have ever held a bicycle wheel by its axle and given it a spin or played with a spinning top toy, you have experienced the momentum of a flywheel. Momentum depends on speed and mass. A bowling ball is very massive and thrown sufficiently fast to travel the full length of the lane will carry enough energy to knock over pins. Try bowling with a low-mass tennis ball and you certainly won't get a strike, let alone get any pins to fall. The same is true of spinning objects. If a very massive wheel is spun up to high speed it carries a lot of energy.

Flywheels have been around for centuries, first used in making pottery then regulating the speed of industrial steam and gas-powered engines. In fact, they are still used today in modern four-cycle combustion engines because only one of those four cycles is power. A flywheel keeps the pistons moving up and down to get rid of exhaust gases, draw in new fuel and air, and then compress it before the next power stroke. Without the flywheel, the engine would not have enough momentum to keep moving through those cycles. But a flywheel can also be used on its own to store and then deliver energy.

Flywheels are pollution free and long-lasting, and once spun up, they can deliver power instantly, so they have the potential to act as energy storage devices for intermittent power supplies such as wind and solar. The wheel is driven by an electric motor that gives the wheel its speed. The momentum of a massive wheel will keep it spinning on its own for some time. Then when you need electricity, the motor becomes a generator driven by the spinning wheel. Of course, taking energy from the wheel slows it down, so you have to spin it up again, like recharging a battery. The amount of electricity you get depends on the mass of the wheel and how fast it spins.

The gyrobus used a flywheel to generate electricity and drive the wheels. The bus could run for about two kilometres. It connected to overhead wires at each stop to spin the flywheel back up.

Image courtesy of Smiley.toerist Wikimedia Commons

Flywheel power first came to light in the 1950s in a Swiss zero-emission vehicle called the gyrobus, which held a 1,500-kilogram wheel housed in a low-friction chamber and spun up to 3,000 rpm. A generator attached to the hub of the wheel produced electricity to drive motors on the wheels. At bus stops, a connector reached up to overhead wires to spin the wheel back up again, allowing the bus to run for about 2 kilometres, which was enough to make it between stops. The system worked for a while, but wear and tear on the fly-wheel bearings required high maintenance and the vehicles were eventually deemed unaffordable.[4]

Modern flywheels, made of high-strength composite materials, can achieve much higher rotational speeds and, therefore, are able to store more energy, but adapting them to vehicles is not practical. To replace a 100-kilogram battery that carries 25 kilowatts of energy, a flywheel would have to be 1 metre wide and spin at 37,000 rpm, which is ten times faster than a conventional engine and not a device you want to have in a family car.

Modern flywheel systems use zero-friction magnetic bearings and massive wheels, and spin at roughly 10,000 rpm. They can provide around 2 megawatts of power for about fifteen minutes before wind-ing down. However, the amount of energy stored is proportional to the square of the angular velocity, which means if the flywheel spins twice as fast, you get four times the energy. But massive wheels undergo more stress the faster they spin, and there is a risk they could fly apart. So lighter-mass rotors are spun up to 100,000 rpm and encased in vacuum containers to reduce air friction. These sys-tems are 95 percent efficient, can be recharged thousands of times, and don't degrade over the long haul.[5]

Flywheel technology is currently used as backup power in indus-trial applications as well as on electrical grids to smooth out power fluctuations. They are even used on the International Space Station to keep the complex properly oriented in space. Spinning wheels do not require exotic rare-earth metals to operate and have a higher capacity than lithium-ion batteries. So, if flywheels are so great, why don't we see more of them around?

It all comes down to cost. Flywheels are high-precision instru-ments that need to be perfectly balanced to achieve the incredibly high speeds. But in truth, it is the incredibly low cost of battery technology that has kept flywheels on the fringe. So much effort has been put into mass production of lithium-ion batteries that they are undercutting many other technologies. However, there are hybrid applications where flywheels are used alongside batteries to stabi-lize the power output, so they are not completely off the table.

The power of falling water has been used for centuries to turn wheels that run millstones to grind wheat or corn. On a larger scale, hydroelectric dams raise the level of water so it can run down through penstocks and past turbines to generate electricity. The water behind a dam is an energy storage device that uses the power of gravity. So how about a rechargeable dam?

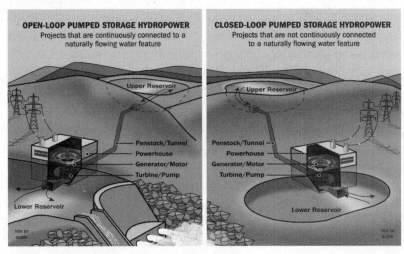

Two examples of pumped hydro energy storage.

Image courtesy of EERE (US Office of Energy Efficiency and Renewable Energy)

In the hills of Vermont in the United States, two reservoirs lie one above the other. A dam holds the upper reservoir in place, forming a lake. When water flows from the upper reservoir to the lower, electricity is produced in the conventional way by running it through turbine generators. But in an innovative new strategy, the flow of water is reversed at night. During times of low demand, excess electricity in the grid is used to pump water from the lower reservoir uphill into the upper lake where it is stored until power is needed. The beauty of this is that the water can be stored as long as necessary. No chemicals are involved, no rare exotic metals. It is a clever idea, but only works where the geography allows.

Feel the Heat

If you have ever made the mistake of walking across a sunny tropical beach in bare feet you know the ability of sand to hold on to a lot of heat. A start-up company in Finland called Polar Night Energy[6] is taking advantage of the heat capacity of sand to provide seasonal energy storage for northern communities. The idea is remarkably simple. Take a large pile of sand, such as leftovers from the construction industry or low-grade sand that is very cheap, and fill either a large, insulated container above ground or a big hole in the ground, and then run pipes that provide heat through the sand. The sand will absorb that heat until it reaches 600°C. Once it gets that hot, the sand will retain that heat, for months if the pile is big enough. This becomes seasonal energy storage that can be connected to local district heating systems to provide heat over the winter when solar energy is at a minimum and energy is in high demand.

The beauty of sand, besides the fact that it is literally dirt cheap, is that you don't have to worry about any leakage, it takes one-fifth the energy to heat up compared to water, and it can be raised to much higher temperatures than water without boiling into steam and creating pressure. The sand just lies there and gets hot, like it does on a beach. It can also be easily scaled up: as the volume of the sand increases, it becomes self-insulating and will retain heat longer. Once a large heat source like that is established, other technologies could be tagged on to provide electricity as well as heat.

One device that could tap into that stored heat and turn it into electricity is the Stirling engine, which has been around as long as the internal combustion engine and is simpler to operate. The engine uses a cylinder and piston, but there is no burning fuel inside. Instead, one end of the cylinder is hot, the other is cool. Air is shuttled back and forth between the two ends, expanding when it gets hot and contracting when it gets cool, providing the pressure to drive the piston. The heat source is on the outside of the cylinder and can come from anywhere, whether that be hot sand, a flame, or solar energy. When the heat engine is attached to a generator, it produces pollution-free electricity.[7]

Another potential application is with the company Ambri, which has developed a liquid metal battery. This highly efficient battery is an alternative to lithium ion and uses molten liquids rather than solid materials to store electricity; however, it needs to operate at around 500°C to keep those metals molten. Perhaps burying the battery in the hot sand might be an ideal partnership.[8]

Australia is also taking a serious look at energy storage in sand. Most of the country is covered in it, and as a desert, there is plenty of solar energy available. An experiment using large parabolic mirrors focused sunlight to make it one thousand times more intense and used it to heat sand. The potential heat storage capacity of 1 hectare of sand 20 metres deep at 600°C is the equivalent of 4,500 tonnes of coal or 10,000 megawatt hours.[9]

The Australians are taking advantage of the fact that sand can flow, as it does in an hourglass, and pumping it through a heat exchanger where it turns water into steam that is run through a turbine to generate electricity.

Feel the Pressure

Air pressure is another way to store energy, using either high-pressure containers or caves. In regions with large underground caverns, or abandoned mines, sections can be sealed off to make them airtight, then large volumes of air can be pumped in under pressure. The biggest problem when compressing air is that it gets hot. Pump up a bicycle tire with a hand pump and you'll feel the tube of the pump get warm. That is not just the sweat of your hand, that is the air inside heating up from compression. As the air heats up it expands, which works against you, so you need to put in more effort to compress it further, which makes more heat, more pressure, etc. The secret is to remove the heat from the air as it is compressed so it will cool and contract, allowing you to stuff more air into your container. That is why dive shops will often immerse scuba tanks in water while they are being filled to get rid of the heat during compression.

On a larger scale, when filling up a cave, the heat created can be captured and stored in some way, for example, in hot water or molten salts. That same heat can then be used to warm the compressed air so it will come out of the container with enough force to drive a turbine and generate electricity. The system can run at very high efficiency because the heat that was captured is recycled back into the system.

Canadian company Hydrostor demonstrated a proof of concept project using large inflatable bags that were submersed deep into Lake Ontario where the weight of the water pressure around the bags provided the force that would release the air when needed.

Since then, the company has moved to using abandoned salt mines to hold compressed air, since salt is naturally impermeable. A facility in Goderich, Ontario, can provide 10 megawatts of power for five hours, enough for two thousand homes.[10]

A project in Australia uses an abandoned zinc mine, where underground storage containers are filled with water. As compressed air is pumped in, the water is forced upwards through pipes to an above-ground reservoir. When the compressed air is needed, the pressure of the water flowing back down squeezes the air back up. It is a variation on pumped hydro where air is the working fluid instead of water.

Taking this idea even further, a company in England stores energy in liquid air. Again, air is compressed and heat is removed, cooling it down, except this time the heat removal is repeated over and over until the air temperature drops to the super-cold temperature of -196°C and turns into a liquid. This is a standard industrial procedure used in all gas liquification processes. The big advantage is that the liquid form of air takes up four hundred times less space than the gas form. A tank of liquid air contains a huge volume of equivalent gas. Return some of the heat that was removed during the liquification process to the liquid, and it will boil back into a gas that comes out under high pressure whenever it is needed. And that gas is just air.

All these energy storage systems take advantage of the law of conservation of energy—energy cannot be created or destroyed, only changed from one form to another—a wonderful gift nature has given us. But there is a corollary to that law: no energy conversion is 100 percent efficient. There is always something lost along the way, usually to waste heat. An important consideration with energy conversions is to count the losses and compare them to the gains. That means making as few energy conversions as possible during the process.

Fossil fuels have allowed us to put up with losses during energy conversions. They have been so cheap and energy dense that we have been able to waste a lot of energy and still have enough left to be useful. A prime example is a coal-fired generating station. A lump of coal contains chemical energy that is converted into thermal energy when it is burned in the boiler. That thermal energy is transferred to water to make steam, which is converted to mechanical energy when it spins the blades of a turbine. The rotational energy of the turbine spins a generator that makes a final conversion into electrical energy. The electricity is then sent out over high-voltage transmission lines where some of the energy is lost due to resistance in the wires. By the time that electricity reaches your home, it represents only about 15 percent of the original chemical energy that was in the coal. In other words, 85 percent was thrown away during all those energy conversions.

Future energy storage systems cannot afford that. If electricity is converted into mechanical energy to pump water, which is raised to a greater height, stored there, then brought back down through mechanical energy and converted back into electricity, the losses during each conversion must be added up to ensure they don't outweigh the gains of storing the energy in the first place.

In the end, energy will come from many different sources and will be highly variable. Energy storage is vital to smooth out the flow of power so excesses can be stored when the wind blows hard or the sun shines brightly, then used to fill in the gaps when the supply drops off. All these different energy storage methods are competing for a very large market but must be proven cost effective on a large scale. Many start-ups face the so-called Valley of Death when good ideas that worked well in laboratory experiments fail to be economical when scaled up to full production. Much of this economic wall has been put up by intense research and development of lithium-ion batteries, which have dropped so low in cost that other forms of energy storage have a hard time keeping up.

But every energy system has its advantages and disadvantages. Batteries degrade over time; water, compressed air, heavy weights, and sand do not. What are the long-term costs? How much power can these systems deliver and for how long? Can they power a region for an hour, a day, or a season? It all comes down to energy management: how we produce it, how much we store, and how much we consume. These are interesting times when many new technologies are emerging on both the production and storage end of our future energy picture.

There is one other consideration: the overall impact of our new technologies on the environment and, in some cases, the social situation.

With all major auto manufacturers bringing electric vehicles into their lineup in an effort to catch up to the popularity of Tesla, there will be a rising demand for metals needed in the manufacture of millions of new batteries. This raises concerns about the environmental and social impact of more intense mining operations.

In a report in the journal *Nature Reviews Materials*, British Earth scientist Richard Herrington points out that by the year 2035, there could be 245 million battery electric vehicles on the road.[11] In addition, there will be a huge demand for stationary batteries for energy storage to compensate for the variable output of clean energy sources such as wind and solar.

As we move towards a carbon-free energy future, Herrington identifies twelve elements that will see increasing demand. These are essential ingredients for batteries and for the manufacture of wind turbines and solar panels. The top three elements—carbon (in the form of graphite), cobalt, and lithium—will need an increase in production of almost 500 percent above current levels.

In some parts of the world, mining does not have a good track record of environmental stewardship. The world's second-largest producer of lithium (after Australia) is Chile, where the metal is obtained from salt brines in the Atacama Desert. Local residents are in an ongoing dispute with the mining companies, claiming the operations, which use a lot of water for extraction, disrupt the water table and delicate ecosystem of the driest desert on Earth.[12]

   More than half of the world's cobalt comes from the Democratic Republic of the Congo, which is known for political unrest and problems with child labour. A large increase in mining operations has the potential to exacerbate environmental and social disruption.

   To avoid some of these issues, the key to meeting this rising demand, Herrington's report says, is effective recycling that will relieve pressure on mining. Industry is already good at recycling some metals, such as aluminum and cobalt, but currently, only 1 percent of the lithium in the global supply chain is from recycled sources. Industry is struggling to develop ways to recycle the lower-value elements of old batteries economically. Using new material is often simply cheaper.

The Industrial Revolution was driven by fossil fuels dug out of the ground. As industry became global, demand for those resources led to conflict, social disruption, and environmental degradation. Now a new green industrial revolution is taking place that depends on different limited resources also buried underground. We need to proceed carefully to avoid those same negative social consequences.

   Energy storage is an important element of our green future, filling in the gaps when the sun doesn't shine, the wind doesn't blow, the tides go slack or when the seas are calm. It will come in many forms and is only a partial solution. There will still be a need for other, more dependable clean sources of energy to ensure the lights will come on whenever we need them.

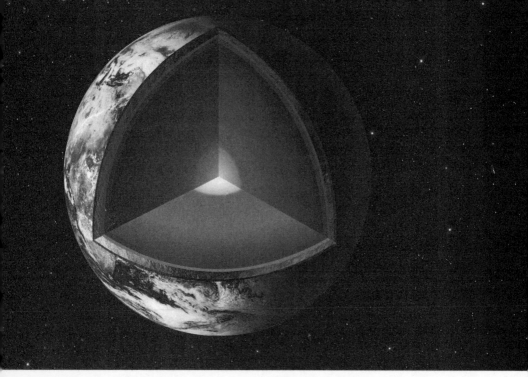

Cross-section of the Earth showing the various layers.

Chapter Seven

Geothermal Energy

Where are the best regions for geothermal and what is the latest technology, both for home use and on a large scale?

The destructive force of volcanic eruptions has claimed human lives for millennia, as the residents of Pompeii and Herculaneum discovered when Italy's Mount Vesuvius explosively erupted in 79 CE. The scene was witnessed from across the Bay of Naples by Pliny the Younger, who described it to historian Tacitus in two letters. In the first, he wrote about the reactions of his uncle, Pliny the Elder:

> My uncle was stationed at Misenum, in active command of the fleet. On 24 August, in the early afternoon, my mother drew his attention to a cloud of unusual size and appearance. He had been

out in the sun, had taken a cold bath, and lunched while lying down, and was then working at his books. He called for his shoes and climbed up to a place which would give him the best view of the phenomenon. It was not clear at that distance from which mountain the cloud was rising (it was afterwards known to be Vesuvius); its general appearance can best be expressed as being like an umbrella pine, for it rose to a great height on a sort of trunk and then split off into branches, I imagine because it was thrust upwards by the first blast and then left unsupported as the pressure subsided, or else it was borne down by its own weight so that it spread out and gradually dispersed. In places it looked white, elsewhere blotched and dirty, according to the amount of soil and ashes it carried with it.

Unfortunately, a trip to rescue victims of the volcano claimed the life of Pliny the Elder, while his nephew, back in Misenum, had to deal with problems of his own on the third day of the eruption:

Ashes were already falling, not as yet very thickly. I looked round: a dense black cloud was coming up behind us, spreading over the earth like a flood. "Let us leave the road while we can still see," I said, "or we shall be knocked down and trampled underfoot in the dark by the crowd behind." We had scarcely sat down to rest when darkness fell, not the dark of a moonless or cloudy night, but as if the lamp had been put out in a closed room.

You could hear the shrieks of women, the wailing of infants, and the shouting of men; some were calling their parents, others their children or their wives, trying to recognize them by their voices. People bewailed their own fate or that of their relatives, and there were some who prayed for death in their terror of dying. Many besought the aid of the gods, but still more imagined there were no gods left, and that the universe was plunged into eternal darkness for evermore.

There were people, too, who added to the real perils by inventing fictitious dangers: some reported that part of Misenum had collapsed or another part was on fire, and though their tales were false they found others to believe them. A gleam of light returned, but we took this to be a warning of the approaching flames rather than daylight. However, the flames remained some distance off; then darkness came on once more and ashes began to fall again, this time in heavy showers. We rose from time to time and shook them off, otherwise we should have been buried and crushed beneath their weight. I could boast that not a groan or cry of fear escaped me in those perils, but I admit that I derived some poor consolation in my mortal lot from the belief that the whole world was dying with me and I with it.[1]

Thankfully, Pliny the Younger survived the eruption of Vesuvius, as did his letters, which provided the only written account of the event. Visitors to the ruins of Pompeii today can see how devastating the power of a volcano can be.

Volcanic eruptions are the most extreme demonstration of the immense amount of energy hidden within the Earth. This deep Earth heat is almost infinitely abundant with more energy stored in the body of our planet than we could possibly consume in a hundred million years. If you could peel away the crust of our planet like the skin of an orange, you would see the mantle, a white-hot crystalline sphere glowing at 900°C that makes up the main body of the Earth. Go deeper and it gets hotter until you reach a liquid outer core at more than 2,000°C and finally a solid inner core made of iron that is a searing 6,000°C. That's hotter than the surface of the sun.[2] The heat is the product of tremendous pressure from the inward pull of gravity along with energy released from naturally radioactive elements within the rock. The Earth has been that hot since it was formed billions of years ago and will continue to be hot for billions more, so there is an unlimited energy supply down there.

Geothermal energy is an attempt to tap into that tremendous resource. But even though it is plentiful, getting to it is not always easy.

You don't have to go far underground in a cavern or mine to feel the warmth in the rocks of the Earth. Tapping into that heat is relatively straightforward, and there are two ways to do it. The easiest method is when the Earth delivers its heat to the surface for us through hot springs. Naturally hot water can be run from the ground through pipes to heat buildings or whole districts of buildings. It can be used for industrial heat, to warm greenhouses, in agricultural drying processes, or simply to enjoy as hot spring swimming pools. The second way to tap into the heat, and use it to generate electricity, involves drilling deeper.

A large-diameter hole is drilled down more than a kilometre to very hot porous rock containing water at 180°C. The hot water rises to the surface under its own pressure and boils into steam, which runs a steam turbine to generate electricity. An alternative method is to drill down to hot rock and pump water down to capture the heat. Geothermal energy is clean, reliable, free of carbon emissions, and available 24/7, which seems ideal for our green energy future.

However, like so many other valuable resources, the Earth does not give up its heat easily or uniformly around the globe.

In the same way that oil is only found in certain areas around the globe, such as central North America or the Middle East, geothermal energy comes to the surface more readily in some places than in others. There are cases such as Canada, where hot springs, geysers, and volcanic regions unfortunately tend to be in mountainous areas, far from most major cities. Heat is not as easy to transport over long distances as electricity. However, there are many places in the world where large populations are close to volcanic regions. Still, geothermal only makes up about 5 percent of our total energy supply. The trick is finding the best spots to get at it then transporting that energy to where it is needed.

The crust of our planet is incredibly thin. If the Earth was an egg, its crust would be thinner than the egg shell. Yet on a human scale, drilling through the crust to get to that very high heat means boring a hole 40 kilometres deep, which is beyond our current technology. The Russians attempted to do this with the Kola Superdeep Borehole on the Kola Peninsula in the High Arctic in the 1970s. But after twenty years of drilling, they reached 12.2 kilometres, a world record but less than a third of the way through the crust.[3] The Americans had Project Mohole, which drilled into a thinner part of the crust on the ocean floor, but that hole only reached 183 metres because the drilling took

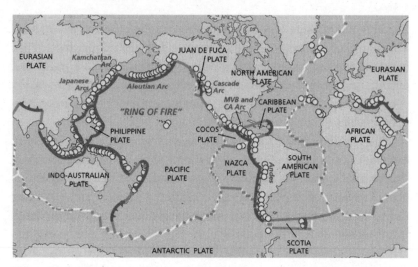

The surface of the Earth is like a cracked egg made up of large plates that move against each other. It is along the boundaries between plates where most earthquake and volcanic activity takes place.

Courtesy of NOAA

place in 3,600 metres of water.[4] Holding a drill ship and the long drill pipe steady in water is extremely difficult at that depth. Since then, the Japanese have taken up the challenge with their M2M, Moho to Mantle project, using the drill ship *Chikyu*. In 2012, the ship set a record by drilling 7,740 metres below sea level and penetrated more than 2 kilometres beneath the sea floor off the coast of Japan. The project was to study the fault zone that led to the devastating underwater earthquake and tsunami the year before.[5]

Thankfully, the crust of our planet is cracked into about a dozen very large pieces, or plates, and along these cracks geothermal energy is more easily accessible. These crustal plates are huge. North and South America are each a single plate, so is the floor of the Pacific Ocean, the continent of Africa, Asia, etc. Because the material under the plates is molten, they are floating, like sheets of ice on a lake during spring breakup. If you were a very large giant, the ground would sink under your feet like the surface of a trampoline as you walked around. In fact, northern Canada was pressed down by the enormous weight of the glaciers during the last ice age and is still slowly rebounding upwards since that ice disappeared about twelve thousand years ago.

Convection currents rising from within the Earth drive the floating plates across the face of the planet like a huge jigsaw puzzle that constantly rearranges the pieces. As the pieces rub and bang into each other along the cracks between them, the heat from within escapes to the surface in the form of volcanoes, hot springs, and earthquakes.

The theory of plate tectonics, as it is known, was developed entirely in the twentieth century. Polar explorer Alfred Wegener first proposed the idea that continents can move when he noted that the east coast of South America and the west coast of Africa have the same shape, as though they once fit together. He introduced the term "continental drift," which initially was not accepted by the geological community. But later work, much of it led by Canadian J. Tuzo Wilson, examined the floor of the Atlantic Ocean and proved that there is a huge crack down the middle known as the Mid-Atlantic Ridge, and that the ocean is getting wider. In fact, about 250 million years ago, South America and Africa were indeed once joined. Wegener has been vindicated.

Plate tectonics is now a fundamental principle in geology that shows how the pieces of the Earth's crust can move against each other in three different ways. They can pull apart, like on either side of the Atlantic Ocean; they can rub alongside each other, like the famous San Andreas Fault in California; or they can collide, which is happening in Italy.

The entire "boot" of Italy is a volcanic mountain range that has been pushed up by the collision of the African and Eurasian plates, which are pushing against each other along the line of the Apennine Mountains. The African Plate, on the west side, is being forced downward, while the Eurasian Plate on the east side rubs over top of it, the way a taller SUV can ride over top of a smaller car during a collision. Those powerful forces within the Earth coming together along that line produce the destructive volcanic eruptions and earthquakes the country is known for. Yet despite the tragic history, those rumblings in the ground also mean there is tremendous heat just below the surface. Ironically, the city of Pompeii, which lay at the base of Mount Vesuvius, had Roman baths that were fed by hot springs from the same volcano that eventually destroyed the city. Little did the people of Pompeii know that the soothing warm waters were coming from the same heat source that would eventually claim their lives. Later, in modern times, Italy was the first country to pioneer a geothermal power plant in 1913 in Larderello in the Tuscany region. Living with a volcano is like having a lion for a pet. It may be tame most of the time, but there is always the potential that it could suddenly turn on you and rip you to pieces.

Another country in a prime location to utilize geothermal energy is Iceland, which uses it for heat and power. The entire island country is one big volcano with its peak sticking above the surface of the Atlantic Ocean. It is the largest mountain in the Mid-Atlantic Ridge, which runs down the centre of the Atlantic from the Arctic all the way to the Horn of Africa, making it the longest mountain range on Earth. Because this line of mountains is almost completely underwater, it was unknown until 1872. This long scar on the face of the Earth is where the North American and Eurasian plates are doing the opposite of what is happening in Italy: they are pulling away from each other. As they separate, hot material wells up from below to fill in the crack and builds a line of mountains in the process. Iceland is the most active part of the chain with eruptions from about thirty active and about a hundred inactive volcanoes scattered around the island. Since the oceanic ridge runs right through the middle of the island, the country is being split apart, making it one of the few places in the world where they *are* making more land!

The people of Iceland, like the ancient citizens of Pompeii, accept the fact that they live on the flanks of an active volcano that occasionally erupts violently. In 2010 an Icelandic eruption threw so much ash into the sky that air travel around Europe was interrupted for weeks, the largest air traffic shutdown since the Second World War. Despite the danger, heat from the Earth's interior is very close to the surface—holes drilled about 3 kilometres deep reach hot rock. Iceland has taken advantage of this abundant energy to generate clean electricity and provide district heating to homes and industry. The country has built six geothermal plants to produce cheap, dependable, clean electricity with no carbon emissions.

That same thermal energy is used to heat buildings by running hot water through insulated pipes. This includes heating high-tech greenhouses, making Iceland a major producer of fresh vegetables, even though it is located close to the Arctic Circle.

Iceland perfectly demonstrates how a country can turn the violence of volcanic eruptions into geothermal energy and do it effectively. Their heat source is not only close to the surface of the Earth but also close to where people live. Heat is difficult to transport over long distances, but since Iceland is a small country, district heating is made easier. One last advantage of geothermal energy is that it doesn't have to compete with fossil fuels. Iceland has no fossil fuels of its own so products like gasoline have to be imported at considerable cost.

This island country is on the right road to a green energy future, including using that abundant electricity to produce hydrogen from water as a fuel for vehicles, which will be discussed in a later chapter. Other parts of the world are not so fortunate.

Canada, for example, is blessed with incredible natural resources, but interestingly, geothermal energy has had a hard time catching on. One reason is the geography, the other is economics.

According to Dr. Catherine Hickson, vice president of Geothermal Canada:

> The problem is that most of Canada is underlain by crystalline bedrock. That's granite and metamorphic rock of the Canadian Shield. It's not nice sandstone that water will percolate through, or limestone, which has caves and things like that. It's tight, tight rock. Unfortunately, that covers 60 percent of Canada.
>
> The Canadian Shield also has quite a low geothermal gradient. It's about 20°C per kilometre. So that's a problem. To be useful for electrical generation, we need about 110 degrees—120 is even better. That means we have to drill 4 to 5 kilometres through hard, dense crystalline rock to get those temperatures.
>
> Because there is no water in these dense hard rocks, you have to create a system of drilling a hole and adding water. The bottom line is: it's very expensive. But it is the focus of a lot of research because of the huge potential.[6]

Canada, like the rest of North America, has most of its volcanic geothermal resources in the mountains in the west. That is where the North American Plate runs into the Pacific Plate, the largest on the planet. Like Italy, that collision has created a line of volcanoes, such as Mount Saint Helens that erupted in 1980, Mount Rainier, Mount Baker, Mount Shasta, and others, stretching all the way from Alaska to Central America. But most of the major population centres and industrial areas are far away in the east, so geothermal cannot support the other side of the continent.

One exception is California, which does have large cities along the coast, so there are forty-three operating geothermal plants there, with the greatest concentration in a hot springs area north of San Francisco. Together they produce 1,800 megawatts, which is only 6 percent of the state's electrical generation. But the third factor, beyond geography and geology, keeping geothermal energy in the background in North America is competition from cheap fossil fuels.

One of the challenges for geothermal energy is copious quantities of natural gas. One of the attributes that geothermal brings to the table over other renewables is thermal energy, heat that is produced alongside electricity. With so many existing pipelines—and by pipelines, I mean community distribution of gas to your homes—it's hard for geothermal to compete with that pipeline infrastructure delivering cheap natural gas.
Dr. Catherine Hickson

While geothermal energy offers a clean, reliable alternative to fossil fuels, as long as those fossil fuels remain cheap, from an economic point of view, geothermal is an expensive alternative.

However, that is not the case in other parts of the world, especially in developing countries such as Kenya. It was the first African country to harness geothermal energy, which makes up 38 percent of its energy output, a greater proportion than any other country in the world. This is all due to its position in the East African Great Rift Valley, which runs 7,000 kilometres (4,350 miles) through the eastern portion of the continent. This is another giant crack in the Earth's crust that is slowly widening, providing tremendous opportunities for geothermal energy.

Countries like Kenya don't have the grid power network that we do, the backup of big impoundment hydro that we have, and access to cheap natural gas for power. So for them, it's perfect. They have a fabulous high-temperature resource. We only have a couple of places like that in Canada. Most of ours is in the 60- to 140-degree Celsius range, whereas Kenya has high-temperature steam resources. They can get good prices for their power, and because they are not a huge country, they don't have long transmission issues to deal with. Kenya has been doing amazing things in the last ten years in geothermal.
Dr. Catherine Hickson

Other countries along the same rift valley, such as Ethiopia, are following suit.

Since geothermal energy works best along the cracks in the Earth's "eggshell," the biggest crack of all is the Ring of Fire in the Pacific Ocean, a huge circular series of subduction zones where plates run into each other. This is the outer boundary of the Pacific Plate. Around the edge of the plate is a huge horseshoe-shaped ring of active fault zones and volcanoes that run from Alaska down the west coasts of North and South America, then pick up again in New Zealand, up through Tonga, Indonesia, the Philippines, Japan, and into Russia. This is where the geothermal heat is closest to the surface and offers these countries their own source of clean energy that pumps income right back into their communities without relying on imports. Geothermal represents only about 1 percent of world energy supply, but that is changing in these active regions with access to geothermal heat.

According to the International Energy Agency:

Geothermal capacity is set to grow 28 percent, or 4GW, to reach just over 17GW by 2023 as projects in nearly 30 countries come online. 70 percent of this growth is in developing countries and emerging economies. The Asia-Pacific region (excluding China) has the largest growth, at 2GW, over the forecast period. Indonesia's expansion is the strongest, propelled by abundant geothermal resource availability and a strong project pipeline in the construction phase supported by government policies. Kenya, the Philippines, and Turkey follow, responsible for 30 percent of additions. [7]

In regions where the Earth is not offering up its heat freely, there are attempts to get at it through a process known as enhanced geothermal. Water is pumped down into deep geological layers under pressure where it forces the rock to crack, allowing more water to be pumped in until an artificial aquifer, or underground pool, is formed. This is similar to fracking, where water and a cocktail of other ingredients are used to create underground cracks that release natural gas.

The fear with enhanced geothermal is that the underground water could act as a lubricant and trigger local earthquakes or slumping of the land above, a phenomenon that has been seen with fracking. Therefore, this technique has to be done in regions where the rock is fairly stable.

A strange juxtaposition exists between the cheap natural gas industry and clean geothermal. One is holding the other back. However, in the future, as fossil fuels are phased out, workers in that industry will become unemployed. But geothermal also involves deep drilling, so would it be possible for geothermal to step in—and even take advantage of gas wells that have already been drilled to extract thermal energy from them?

It is hard to get people to understand what is different about geothermal. The surface facilities, the pad that the drill rig sits on, the access roads, storage areas—that all can be reused.

The wells, though, are quite different. We build geothermal to last decades, fifty years or more. To do that, we drill them to a different standard than what you drill oil and gas wells to. When you flow a well, those fluids that you are bringing out of the ground are actually quite nasty. They can be very high in salt—some of them are sour, which means they have H_2S or sour gas, and that can be very corrosive. The salts can precipitate onto the linings of the well. In the oil and gas sector, they are not terribly worried about it. The value of the oil and gas they are extracting allows them to only use the well for a few years in order to recover the drilling costs and make a profit.

Oil wells are also too small in diameter. When we are talking about temperature, let's say 120°C, we are hoping to make 10 megawatts of power. That means we have to flow 300 litres per second of fluid. A typical oil and gas well might flow 15 litres per second. And even 15 litres per second is a lot of water that they have to get rid of after the separation of the hydrocarbons from the water. So they target wells that don't flow a lot of water.

Our geothermal wells are significantly larger in diameter, with much bigger pumps than you would put into an oil and gas well.

Dr. Catherine Hickson

While it might not be an easy switchover from fossil fuel drilling to geothermal energy drilling, much of the technology and expertise can be re-purposed in the future when fossil fuel prices rise, or a higher price is placed on carbon. And speaking of carbon, according to Dr. Dickson, geothermal could also play a role in getting rid of it.

The other thing to add value to geothermal, besides electricity and heat, and make it more trendy, is to add CO_2 sequestration. We've got to drill these wells anyway. We're going to be injecting water, so why not add CO_2?

I think this is going to be a bit of a game changer for geothermal, where you need another revenue stream and you need another angle to capture people's imagination. Developments are likely in areas where significant amounts of CO_2 are being produced by other industries, like hydrogen, for example, so why not sequester their carbon?

Like other forms of energy, geothermal can also work on a smaller scale, such as household heat pumps, which have been around for decades. Heat pumps are basically air conditioners that can work in two directions, either removing heat from a home and dumping it outside, or gathering heat from outside and bringing it in, even when the outside temperature is cold.

How does a heat pump work?

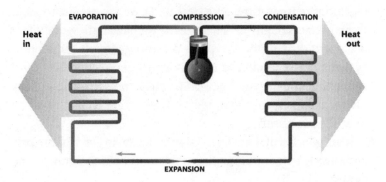

Simple schematic of a heat pump.   © designua/Adobe Stock

All refrigeration or air conditioning equipment uses a working fluid that has a very low boiling point. That means it can be turned from a liquid to a gas at around room temperature. Freons and halocarbons are most commonly used. Previously it was CFCs, or chlorofluorohydrocarbons, but they turned out to be harmful to the Earth's ozone layer, so they were banned in the 1980s. Today's units use modified versions, a laundry list of chemicals that are not harmful, including simple carbon dioxide.

The root of the system is the principle that when a gas is squeezed, or compressed, it releases heat. If that heat is then taken away, and the pressure on the gas released, or allowed to expand, it wants to get that heat back. But since the heat has been taken away, the gas cools. The process can be repeated over and over in a closed loop.

In an air conditioner, the heat from the room is absorbed by the liquid, which boils into a gas. The gas is compressed back into liquid and the heat produced is given off to the outside air. The cooled liquid returns into the house to cool the room. A heat pump can do that, but also run in reverse, bringing heat from the outside in. That may seem odd if it is cold outside, but as long as the refrigerant liquid is colder than the outside air, it will absorb heat and bring it inside to heat the home.

Heat pumps cannot completely replace furnaces because they lose their efficiency when the temperature drops below -10°C, but they can cut heating bills by about 50 percent, which also means reducing greenhouse emissions if the home is heated by fossil fuels.

An alternative is to use the heat of the Earth to run a heat pump instead of the outside air to heat or cool. Again, the principle is the same, but this time, pipes must be laid underground, either horizontally just below the surface or in a vertical hole. This adds considerably to the cost of installation. The advantage here is that the ground is always about the same temperature regardless of the air outside, so it will be cooler than the air in summer and if the hole is below the frost line, it will always be above freezing in the winter. Running fluids through the pipes is a way to stabilize the temperature in a building throughout the seasons.

Many new high-rise condo developments are being built with central heat pumps in mind, where pipes are laid into the ground under the foundation before the building is erected. The same could be done for industrial parks, where utilizing heat from the ground and distributed heating can be incorporated into the overall design.

Geothermal energy has shown limited growth because of the high installation costs compared to the relatively low cost of fossil fuels such as natural gas. Heat pumps have always been considered a companion to traditional energy sources. However, there is a growing trend to include ground piping in new developments before the foundations are poured to lower down-the-road, ongoing costs of the heating and cooling.

If you have an opportunity to visit an active volcano, whether it be in Iceland, Italy, the Big Island of Hawaii, Indonesia, or elsewhere, you will find it a powerful, profound experience.

During a trip to Costa Rica in Central America, which is part of the Ring of Fire, I had the opportunity to see a spectacular fire show put on by Arenal, the country's most famous volcano. I was fortunate to be there in the 1990s when it was active. In 2010, it entered a quiet phase but could reawaken at any time.

Coming upon the volcano was an accident, thanks to getting lost and looking for a hotel after dark. After running into a dead end on an unlit country road and wondering where to go next, I looked up to see strange lights moving above in the black sky like some kind of

UFO. There were twin beams pointing upwards and moving errati-
cally. Looking more closely, I could see two red lights following along.
It was a vehicle climbing a steep hill in the darkness. Curious, and
thinking perhaps they were heading up to a mountain resort, I retraced
our route and found a narrow dirt road leading in that direction.

The road immediately began to climb steeply, winding between
large rocks strewn along either side. My vehicle was not a four-wheel
drive, but as long as it managed the incline, I continued on. Eventually the road curved around the downslope side of a house-sized
boulder and wound up onto the flat topside where, much to my surprise, there were about a dozen parked vehicles all facing the same
direction. At first I thought it was a local lovers' lane, but as my head-
lights swept across the group, I could see people sitting on the hoods
of their cars reclining back against the windshields looking up.
A few of them waved at me to turn off my lights.

Finding a parking spot, I got out of the car and looked up to see
what had everyone's attention. Slowly, as my eyes adjusted to the
dark, I could see a red glow high above us. It was the fiery mouth of
the volcano spewing red-hot ash into the sky. Glowing cinders shot
straight up into the blackness then arched down in a fountain of fire.
Along with them were hot boulders clattering down the steep slopes
of the peak.

Then came the sound, a raspy, huffing noise like a giant steam
locomotive churning hard inside the mountain. HUH, HUH . . . HUH!
The volcano was coughing and spitting, a dragon in restless sleep
that could wake up in anger at any moment. I was astounded that
the local people felt safe on the very flanks of this fiery beast. Everyone was quiet, awed by the mighty power of the Earth so close, so
deadly, yet so beautiful. It was a truly humbling experience to get a
glimpse into the immense energy that lies beneath our feet.

Like the simmering heat within a volcano, geothermal energy is
a sleeping giant waiting for the right time to erupt onto the clean
energy scene in a big way.

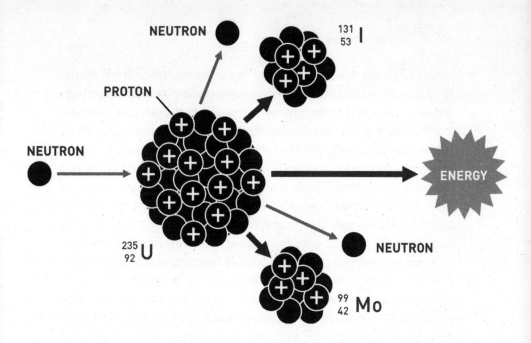

NEUTRON

PROTON

$^{131}_{53}$ I

NEUTRON

ENERGY

NEUTRON

$^{235}_{92}$ U

$^{99}_{42}$ Mo

Uranium atom splitting, neutrons hitting other atoms and splitting in a chain reaction.

Image courtesy of Canadian Nuclear Association

Chapter Eight

Small Nuclear

Taking another look at nuclear energy by thinking small, modular reactors.

A journalist colleague at the CBC travelled to the North Pole aboard a Canadian icebreaker. After crunching through ice for days, they were surprised to be met at the Pole by a much larger, nuclear-powered Russian icebreaker. The Russian captain invited the Canadians over for dinner and a tour of their impressive vessel. The captain asked his Canadian counterpart how much diesel fuel he burned reaching the Pole. When the Canadian calculated the number of tons of fuel used by his ship, the Russian chuckled and replied, "I think we burned a few grams of uranium."

Albert Einstein was well aware of the powerful forces held within the nucleus of an atom. That's why he realized it would make a fearsome weapon if that energy was released all at once. He was concerned that Nazi Germany was on the way to developing a nuclear bomb, which prompted him to write a letter to US President Eisenhower to encourage the Americans to develop one first. Einstein was not directly involved in the Manhattan Project that did lead to the first atomic weapons, but after he saw how one single bomb could completely devastate a city, and later found out that the Germans were nowhere near developing the bomb, he regretted his decision to write the letter and spent his later years denouncing nuclear proliferation. Among the quotes attributed to him is this one from 1945: "The release of atomic power has changed everything except our way of thinking . . . the solution to this problem lies in the heart of mankind. If only I had known, I should have become a watchmaker."

It is unfortunate that the mushroom cloud and the terrible destruction it represents is often associated with nuclear power. True, it was first used in the atom bomb. But there is another more beneficial way to release the forces within the atom in a slower, controlled way to provide reliable emissions-free energy. And that is through a nuclear reactor.

Professor Einstein's famous formula, $E = mC^2$, says that mass can be converted into energy—a lot of energy. The formula translates as, Energy (E) is equal to mass (m) multiplied by the speed of light (C) squared. The speed of light is a large number: 299,792,458 kilometres per second. So even a small amount of mass multiplied by that big number and squared delivers a huge amount of energy. That is the fundamental principle behind nuclear energy, where the forces that hold protons and neutrons together in the nucleus of an atom are broken, and enormous amounts of energy are released. Einstein considered mass simply as energy waiting to be set free.

To put that into perspective, an average adult has a mass of 62 kilograms. If the mass of a human body was turned entirely into energy, it would release 560 billion megajoules of energy, which is thousands of times more than the power of the first atomic bomb. Thankfully, the immense power of nuclear weapons became a deterrent to their further use. While superpowers armed themselves, neither side wanted to use them because retaliation would be so severe that both sides would destroy each other. The concept, known as mutually assured destruction, MAD, became the basis of the Cold War.

For decades, peacetime nuclear energy has provided a backbone of baseline power for Canada, the United States, France, and Japan, among other countries. Now, after more than half a century of nuclear power, a different kind of cloud hangs over the industry in the form of cost overruns, long development times, three major accidents, and a negative public image. Still, it is the most energy-dense fuel of all, with the potential to provide plentiful, 24/7, emissions-free power.

Today, as we look to clean alternative sources of energy, the hard reality is that wind, solar, and geothermal cannot meet all our world energy demand. It is simply not possible to cover all the land with solar farms and wind turbines. And those resources are not evenly spread around the world. Some areas get more sun than others, some areas are windier, and locations are not always near major urban centres. Currently, fossil fuels make up more than 85 percent of our energy sources with only 11 percent carried by renewables (hydro-power, wind, solar, geothermal, wave, tidal, and biofuels). By 2050, fossil fuels will still dominate. Many scientists believe we need more nuclear to replace those remaining fossil fuels, and it needs to be approached differently than in the past.

Nuclear reactors capture the energy of the atom without an explosion by slowing the reaction down over a long period of time and using the heat to generate electricity without emitting carbon into the atmosphere. And all of this is done using a remarkably small amount of fuel.

The natural uranium used in many reactors is in the form of pellets, each about the size of an AA battery and weighing only 20 grams. The energy in one pellet is equal to 400 kilograms of coal, 410 litres of oil, or 350 cubic metres of natural gas. Fewer than ten pellets could power an average home for a year. Imagine powering your home with ten flashlight batteries.

Nuclear generating stations have been megaprojects costing billions of dollars to build, years to construct, and, when things go

THE POWER OF URANIUM

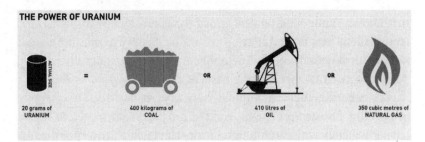

| 20 grams of URANIUM | 400 kilograms of COAL | 410 litres of OIL | 350 cubic metres of NATURAL GAS |

Uranium is incredibly energy dense compared to fossil fuels.

Image courtesy of Canadian Nuclear Association

wrong, billions more dollars to clean up. This has pretty much stalled the nuclear industry, with no new reactors for more than thirty years in the United States, a country that accounts for 30 percent of world nuclear power.

Now, in the face of climate change, an urgent need to replace fossil fuels, and the need for backup baseline power when wind and solar are not available, scientists and engineers are rethinking the approach to nuclear power. Reducing the size of reactors from football-stadium-size to units that can be carried on a flatbed truck, manufactured assembly-line style, and buried underground could make nuclear energy safer and more economically feasible. Those units are called SMRs, Small Modular Reactors.

All nuclear reactors draw their power from the nucleus of an atom. Think of the atom as an incredibly small, incredibly hard nut at the centre surrounded by a cloud of even smaller electrons dancing around it like sparks around the head of a sparkler. To put it in perspective, the simplest and most abundant element in the universe, hydrogen, has one positively charged proton in the nucleus and one negatively charged electron swirling around it. If the proton was the size of a golf ball, the electron, still barely visible to the human eye, would be a kilometre and a half away. So, atoms are really mostly empty space. Yet within all that nothingness is tremendous energy.

All the chemical reactions we see in everyday life, such as striking a match to make fire, involve those outer electrons being ripped off and exchanged with other types of atoms, a process that can release heat and light. That's chemistry, and we drove the Industrial Revolution by simply burning things. But the forces holding electrons to the nucleus are not nearly as strong as those within the nucleus itself. Nuclear energy goes deeper by starting with heavier elements like uranium, which has hundreds of particles in the nucleus, then breaking them apart like a rack of billiard balls on a table. That releases far more energy than you see in the flame of a match. Again, by comparison, just 200 grams of uranium fuel is the equivalent of 4 tonnes of coal.

A nucleus has two basic units, a proton, which is electrically charged, and a neutron, which has no charge. (The only exception is hydrogen, which is just a single proton and one electron going around it.) Usually there is an equal number of both, and they are held together by very strong nuclear forces. The heavier the element is, the greater the number of protons and neutrons. Since protons carry a charge, they are easy to count, so the number of protons in a nucleus gives the element its atomic number. Oxygen has 8, iron has 26. The number of protons is matched by an equal number of electrons that form a multi-layered cloud around the outside. Then you get into the heavyweights such as uranium, which has 92 protons. The advantage to these massive elements is that extra neutrons can be packed in, like styrofoam beads around a delicate object in a shipping box. These are the isotopes. In the case of uranium, one isotope contains an extra 51 neutrons to make uranium-235, which is the fuel for nuclear reactors.

So how do you go about releasing those powerful forces that hold the nucleus together? You hit it with pieces from another nucleus.

It turns out that heavy elements like uranium-235 or plutonium-239 provide the cue balls in the form of neutrons that work for us to break apart other nuclei. These big isotope atoms have so many extra neutrons that they occasionally shoot one off at high speed. That is natural radioactive decay. If there happens to be another atom nearby and it is struck by that stray neutron in just the right way, two things can happen. The second atom could absorb the neutron and become heavier, say uranium-236. Or, if the neutron is slowed down before it strikes the second atom, rather than being absorbed, it will crack the atom's nucleus, creating a fissure that breaks it into smaller pieces. This is fission. That split releases energy in the form of heat and radiation.

A cocktail of by-products is produced by the split, such as cesium, rubidium, barium, and krypton (I guess Superman should stay away from nuclear reactors!).

Each split of a uranium atom releases a lot of heat as well as two more neutrons that can go on and strike two other atoms causing them to split and so on, with the number of splits doubling every time. If you cram a large number of these atoms together into a small enough space, the so-called critical mass, a rapidly escalating chain reaction can occur. If you do not control that reaction and let it run away on its own, before you know it—BOOM!—so much energy is released in such a short time that an entire city is destroyed in seconds. Only 64 kilograms of uranium-235 was used in the bomb that destroyed Hiroshima. A more powerful modern nuclear weapon uses highly concentrated plutonium smaller in size than a grapefruit.

To be clear, a nuclear power generating station cannot explode like a nuclear bomb. The uranium in a reactor is only about 2 to 5 percent fissile material. Bombs use highly enriched uranium that has been concentrated to around 95 percent or higher. A runaway reaction in a power station can cause the fuel to overheat and melt, which has happened in the past, but it will not detonate. There are simply not enough reactions taking place. So don't worry about a nuclear reactor going up in a huge mushroom cloud. It won't happen, although there are other side effects that we will discuss.

Nuclear reactions in generating stations are carefully controlled so that heat is released evenly over a much longer time and used to turn water into steam, which is run through turbines to generate electricity. The secret to keeping a nuclear reaction running for a prolonged period is to slow down those neutrons that are given off, the cue balls that cause the splitting.

How a Nuclear Reactor Works

"Nuclear power is one hell of a way to boil water."

Albert Einstein

A nuclear reactor is basically a giant kettle that boils water to make steam. Except rather than a flame burning under the kettle, the heat comes from reactions taking place within the reactor core. The secret is to keep the reactions at just the right pace so heat is produced at a steady rate, rather than all at once. To do that, there are three basic components: fuel bundles, cooling water, and control rods.

Uranium fuel is ground into a powder and pressed into cylindrical pellets. These pellets are loaded into half-metre-long zirconium alloy tubes that are grouped into bundles that look like a bunch of drinking straws tied together side by side. The assembled bundle is about the size of a fire log and contains enough energy to power one hundred homes for eighteen months or more, depending on the concentration of the fuel.

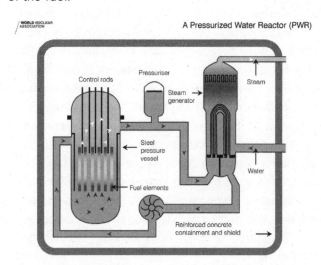

A Pressurized Water Reactor (PWR)

WORLD NUCLEAR ASSOCIATION

Control rods

Pressuriser

Steam → generator

Steam

Steel pressure vessel

Fuel elements

Water

Reinforced concrete containment and shield →

Diagram of a nuclear reactor showing the core, circulating water, turbines, and generators.

Image Courtesy of World Nuclear Association

The bundles are fed into channels inside the reactor core. For a 790-kilowatt CANDU reactor, there are 5,760 bundles holding 5 million fuel pellets that can provide electricity for half a million families.

A nuclear core is submerged underwater for several reasons. Water keeps the reactor at the right temperature and acts as a natural absorber of radiation. But this second component in the reactor also plays another, counterintuitive, role: water helps the reaction along by slowing down neutrons.

It turns out that the fast neutrons emitted naturally by the uranium do not cause the splitting of other atoms. They are either absorbed or just fly off. But as some neutrons pass through water, they are slowed down to the right speed to cause fission. You would think that the harder you hit something, the more it would blow apart—again, like a cue ball hitting a rack of billiard balls really hard to make them scatter as much as possible across the table. But that is not the case with neutrons. It is the slower neutrons that have a greater chance of hitting other atoms and causing the chain reaction. Water is a moderator that slows the neutrons for the nuclear reaction to take place.

The third component, control rods, are usually made of boron or some other material that absorbs neutrons like a sponge. The rods are slid between the fuel bundles to control the speed of the reactions, or to absorb all the neutrons and totally shut down the reactor. It's like removing the cue ball from a billiards table. There is no game without it.

Most reactor designs place the core inside a stainless-steel vessel resembling a huge thermos bottle, which is inside a concrete containment building with walls more than a metre thick to contain any accident that happens on the inside, and prevent any airplanes crashing through from the outside.

Once the reactor is operating at a constant temperature, it is simply a heat source to boil water. The water that runs through the core is kept in a closed loop so it doesn't escape; however, through a heat exchanger, it gives its heat to a separate loop of water on the outside that becomes steam to run the turbines.

There are variations in design, but the principle is the same.

Despite its negative public image, nuclear power has the best safety record of any energy production, which may come as a surprise to some. Just the word "nuclear" can strike fear in the hearts of many. This is not to say it is perfectly safe. It's a bit like a pit bull: the chances of being bitten by one are extremely small, but if you are bitten, it can be very serious because the dog has such a powerful bite. Nuclear reactors have sat quietly around the world churning out electricity for decades without major incidents, but on five occasions, things went very bad very quickly. Three of them gained world attention, Three Mile Island in the United States, Fukushima in Japan, and the Chernobyl incident in Ukraine in the former Soviet Union. The other two lesser known accidents were much earlier in 1957 in Great Britain and the Soviet Union. None of these were nuclear explosions, but meltdowns, which still have serious outcomes.

A meltdown is usually caused by a loss of cooling water, so the core becomes so hot it literally melts. In 1979, a broken valve in the Three Mile Island reactor in Pennsylvania allowed cooling water to drain out of the reactor. Control rods shut the reactor down within ten seconds, but residual heat in the core began to build up. A confusing control panel led controllers to believe the water was flowing in the opposite direction, so they let more out until the levels got so low the remaining water boiled off into steam, which does not absorb heat as well as liquid water. About half of the core melted and settled into the bottom of the steel container. Some of that steam escaped into the air, but most of the radioactive elements were filtered out before they left the building.

By pure coincidence, the Hollywood movie *The China Syndrome* premiered just twelve days before the accident. The movie portrayed almost the exact same scenario as the Three Mile Island accident, but provoked fear that the core could melt through the floor of the building into the Earth and all the way to China. That didn't happen. The melted core was contained within the reactor vessel and, years later, completely removed and disposed of. No one was injured at Three Mile Island, and the levels of radiation released were the equivalent of one chest X-ray. In an epidemiological study, there were no significant increases in cancer rates in the years since, and there have been no deaths directly related to Three Mile Island. Still, in the public mind, it was a total disaster.[1]

Japan suffered a similar fate in 2011 when an underwater earthquake occurred just off the coast. Tremors rumbling through the ground triggered the release of control rods into the reactors at the Fukushima Daiichi plant, shutting down reactions. In a strange twist of irony, the power went out, so a facility designed to make electricity had no electricity to keep the cooling water running through the cores. Backup diesel generators fired up to run pumps, all systems were working properly, and the system was under control. Then a tsunami washed over a seawall surrounding the plant and onto those generators that were unfortunately located at ground level. Had the plant been designed with the generators on the roof of the building, disaster would have been averted.

With the pumps out of action, temperatures rose to 2,300°C, and three of the six reactor cores melted down. (Two reactors were in cold shutdown before the tsunami and were not affected.) Fire trucks were brought in to spray sea water onto the cores, and some of that contaminated water was released into the ocean. Fishing was shut down offshore of the reactor and ocean currents spread the radioactive water across the Pacific. There was great fear that the entire west coast of Canada and the United States would be contaminated for thousands of years, but when the ocean water was tested, the dilution effect of the vast Pacific ocean had reduced radiation levels far below the natural levels found in seawater.

Melted core material dripped down to the bottom of the steel containment chamber, melted through the bottom and into the outer steel and concrete container. None is believed to have escaped into the environment. Water remaining in the core boiled into steam, which was diverted into containment units to be cooled, but eventually these chambers became heated and pressurized as well. With pressure rising to dangerous levels, operators opened valves to release the steam into the air. This steam contained radioactive by-products.

Video footage of the accident shows explosions blowing the roofs off buildings, but these were caused by hydrogen gas that was created from water hitting the hot cores, not the nuclear reactors blowing up. The hydrogen leaked out, accumulated at the top of the outer buildings, combined with air, and exploded. The containment structures around the nuclear cores were not breached by these explosions and remain on-site to this day. In the years following the disaster, circulation pumps have been restored to cool the damaged reactors and the site has been cleared. The new challenges will be dealing with contaminated water held in storage tanks and removing the melted core material, which may take decades.

Thanks to quick action on the part of plant operators who fought to regain control of the reactors, there were no deaths directly related to the meltdown and later studies found no elevated incidence of cancer related to radiation poisoning. Sadly, many did lose their lives during the evacuation of the area due to fears over the reactor meltdown and the devastating effects of the tsunami.[2]

The absolute worst nuclear accident was the tragic disaster at the Chernobyl site on April 26, 1986, in Ukraine, where the core was blown to bits—again, not a nuclear explosion. Had it been a nuclear detonation, there would be no reactor building left. The entire area, including the town, would have been levelled and turned to glass.[3] Still, the explosion that did occur blew radioactive material high into the atmosphere where it was scattered across Europe and eventually around the globe.

The older-style reactor designed and built in the Soviet Union is called an RBMK, which translated from Russian stands for "high-power channel-type reactor." It used solid graphite as a moderator instead of water, although water was still used for cooling. Unfortunately, graphite burns. In another twist of irony, a safety test was being conducted to see if the water pumps could be kept running in case of an emergency shutdown. The reactor was scheduled to be shut down anyway for an annual exchange of fuel rods.

The engineers wanted to conduct the test when the reactor power was reduced to about half. But that day, the reactor had been running at full power for some time, which caused a buildup of xenon gas, one of the by-products of fission. Xenon is known as a reactor poison because it absorbs neutrons, stealing them from taking part in the reaction. Normally, this is not a problem as the reactor can be adjusted by lifting control rods to compensate for the loss. But as the reactor power was brought down for the test, this extra absorption by the xenon caused the power to suddenly plummet well below the halfway level at which the engineers wanted to do their test. At this point, according to their own operating regulations, the reactor should have been allowed to power down completely and left for a few days while the xenon buildup was dealt with. And this is where human error came into play.

That night, so the test could go ahead, a decision was made to bring the power back up to half by removing some control rods. The xenon held the power down. More control rods were pulled out. Remember, control rods slow the reaction down so when they are removed the reaction is free to continue. The graphite is the moderator that assists the nuclear reaction, and it is always there. As more control rods were removed, the reactor suddenly took off out of control, and in a matter of seconds, temperatures quickly rose to extreme levels, more than ten times higher than the reactor was designed for.

There are several theories about what happened next. One is that the cooling water turned to steam within the reactor core and blew it apart in a steam explosion. Another suggests the temperature became so high the water was broken down to its two elements, hydrogen and oxygen, which is very explosive rocket fuel. By this time, the graphite was glowing red hot like charcoal briquets, which, combined with the dangerous gases, triggered a chemical explosion. Perhaps it was a combination, but the explosion shattered the reactor, blew the roof off the building, and set the graphite on fire. Tonnes of radioactive debris were thrown thousands of metres into the air where it was scattered far and wide on the wind.

Thirty-four men died from the blast and acute radiation exposure; another sixty, the firefighters and workers sent in to put out the fire, died in the decades since. The total number of deaths from exposure to the radioactive fallout in the surrounding area varies from four thousand to sixty thousand, depending on who you ask. The most common cause of death is from thyroid cancer caused by the body absorbing radioactive iodine, one of the by-products of the nuclear reactions. But iodine also occurs naturally, and the body needs it.

So thyroid cancer is common, which makes it difficult in the long term to tease out which cancers were caused by nuclear fallout and which would have happened anyway. In any case, Chernobyl was a tragic case of human error and the failure of an old reactor design that is not used in other parts of the world.[4]

The other two accidents[5] happened during the cold war when reactors were hastily built to produce plutonium for nuclear weapons. On October 10, 1957, fire broke out at the Windscale Cumberland (now Sellafield Cumbria) unit one and raged for three days. The accident released radioactive contamination across Europe and is believed to have been responsible for several hundred deaths by cancer. It was the worst nuclear incident in the UK.

That same year an explosion took place at the secretive Myak plant near the Russian town of Kyshtym. A faulty cooling system allowed temperatures to rise leading to the explosion. Ten thousand people were evacuated from the local area. The number of deaths is unknown because it was a secret military operation.

These accidents were very serious, but the only major failures among more than 400 nuclear reactors worldwide that have been operated since the first successful controlled nuclear chain reaction was performed in 1942 by Enrico Fermi under an abandoned football stadium at the University of Chicago. Since then, the average number of deaths from nuclear power operations is one person every fourteen years, and even if you add the highest estimate from Chernobyl, it pales compared to the estimated four to eight million deaths every year from burning coal, oil, and natural gas.[6] In addition, there are the environmental catastrophes from oil tanker spills, oil well spills, train derailments, pipeline ruptures, and of course the warming effect on the climate. That is not to say nuclear power is totally safe, but compared to other forms of energy production, it has the best safety record.

Even with strict safety regulations and technologies in place to prevent accidents, there is still the issue of what to do with the radioactive waste products that come out of nuclear reactors. When the uranium has done its job generating power, less than 5 percent of it has actually been used. The remaining by-products are still radioactive and still producing heat, which make them impossible for humans to handle. That is why all waste from nuclear reactors is stored on-site under strict supervision.

Fortunately, nuclear fuel is so energy dense that, relatively speaking, there isn't very much of it. The total quantity of spent fuel bundles in Canada, which operates nineteen CANDU reactors, would fill about seven hockey rinks to the height of the boards. That's after more than fifty years of operation. Compare that to the billions of tons of combustion products that are released into the air every year from burning fossil fuels.

Above-ground storage is one of three methods to store nuclear waste.

Image courtesy of Nuclear Waste Management Organization

Of course, nuclear waste is not stored in hockey rinks, but it is temporarily stored in what look like oversized Olympic swimming pools. Water is a natural absorber of radiation, so spent fuel bundles are robotically moved from the reactor into these pools where they are stored for at least ten years. Visiting one of these pools at the Bruce Power Generating Station in Ontario, the largest in North America and second largest in the world, is a somewhat eerie experience. The first thing you notice is that the pool has no cover. The crystal-clear water is almost inviting you to dive in. That's how good the water is as a shield, although I wouldn't recommend going for a swim. Resting on the bottom of the pool are racks of fuel rods that look like a warehouse at a metal factory. But as you gaze towards the opposite end of the pool, closer to the reactor area, a beautiful blue glow appears, as though some kind of ghostly spirit is living in the depths of the pool. It is another case where nature makes something beautiful out of something deadly toxic, like the rainbow colours produced by an oil slick floating on water.

The blue glow around nuclear fuel is known as Cherenkov radiation, named after Soviet physicist Pavel Cherenkov, who observed the phenomenon in 1934 when a bottle of water exposed to radiation glowed blue. He and his colleagues Ilya Frank and Igor Tamm went on to win the Nobel Prize in 1958 for their experiments. The haunting blue glow, about the colour of a clear blue sky, seems to come from everywhere at once and has no definable edge. It just fades into the water like a mist. It's a fascinating phenomenon that is the product of particles travelling faster than the speed of light through water, which may seem to go against one of the fundamental laws of physics, but in water it can happen.

It is true that in space, in a vacuum, light travels at 300,000 kilometres per second, which is a universal speed limit. Nothing can go faster than that. At least we haven't found anything yet that breaks the light barrier. But when light passes through something like air, glass, or water, it slows down. So light speed in water is only three-quarters what it is in space. But the particles that come out of fission reactions are moving at 99.9 percent the true speed of light, so in water, they are the equivalent of supersonic jets breaking the sound barrier. In the same way that air cannot get out of the way fast enough when a jet passes through, producing a shock wave and sonic boom, the atoms in the water are disturbed by particles, particularly electrons, that are passing through faster than the atoms can handle. This sets up a cone of disturbance behind the electron, where the water atoms are shaken out of place, giving off light in the process. At least that's the simplified version. To describe it fully requires a complicated set of equations involving relativity and quantum theory that was figured out by Albert Einstein.

The blue glow of Cherenkov radiation is a sign that fission reactions are taking place. Engineers use that glow to determine how active the spent fuel is. Over a period of a decade, the fuel rods cool down, the glow disappears, and the rods are shifted to the far end of the pool where they can be moved into specialized containers for above-ground storage. Future plans call for storing nuclear waste in underground facilities, which presents another interesting problem. If these storage facilities are to hold hazardous waste for thousands of years, how do we keep people away from them in the future? We have never built structures to last that long. The pyramids of Egypt are only about five thousand years old and they have been opened and explored for looting and scientific curiosity.

Example of a nuclear waste site warning sign. Will this be understood thousands of years from now?

Image courtesy of Wikimedia Commons

Will these nuclear waste sites have to be marked in some way with "Keep Out" signs in every language? What if languages do not survive for millennia? Perhaps it's best to leave no marker at all so the hidden vaults will be completely forgotten.

The important point here is that all nuclear waste is under lock and key. Yes, it is hazardous and needs to be cared for over very long periods, but the small amount of waste involved and the fact that it stays in one location makes it much more benign than the billions of tons of waste from burning fossil fuels that is thrown into our air, water, and land.

One possible alternative to reduce the amount of nuclear waste is to substitute the element thorium for uranium. This material is cheaper and more abundant in the Earth and does not produce the same level of radioactive waste. Thorium was considered in the early days of nuclear development but because it does not produce plutonium, which is used in nuclear weapons, was essentially neglected. Uranium became the fuel of choice.

At this point, you might be wondering why a book describing clean technology is spending so much time describing the three worst nuclear disasters and the problem of waste products that last for thousands of years. It is precisely because those events led to nuclear energy falling out of favour in North America and Japan. Public fear, which influences political decisions, has resulted in longer development times and higher costs, including the costs of retrofitting older plants to keep them running longer.

Despite the problems, nuclear energy is considered relatively clean because there are virtually no carbon emissions. But the giant mega-plants are multi-billion-dollar projects that take years to build, must be located away from residential areas—which means more energy lost due to overland transmission through power lines—and no one wants them in their neighbourhood. Yet the amount of power nuclear stations can generate is huge compared to wind and solar farms, and they can keep that power flowing reliably 24/7, so it is hard to write off nuclear energy completely. Rethinking it may be timely.

One approach is to take the same technology used in the mega-

Small modular nuclear reactor on the back of a flatbed truck.

Image courtesy of Nuscale

plants and shrink it down to small modular reactors (SMRs), self-contained units about the size of a school bus that produce less than 100 megawatts of power. Large nuclear power plants can produce more than 4,000 megawatts. The advantage of modular is that if you need more power, just add another module. Small reactors have been in use in one form or another for more than fifty years as research reactors at universities as well as power sources for nuclear submarines, aircraft carriers, and icebreakers. When you think about it, an aircraft carrier is a small floating city with a nuclear reactor right in the middle. There are more than 150 SMR designs worldwide. Some are in commercial use already, some have been physically demonstrated, some are being prototyped, and some are still in the research phase.

Modular means the reactors are all made in the same factory and are identical, which brings down costs considerably. Think of the Model T cars about which Henry Ford once famously said, "They come in any colour as long as it is black." In the past, large-scale nuclear plants were each custom-built to fit the location, which drove costs up into the tens of billions of dollars. Think about how much cheaper it is to buy a new car that comes fully assembled from a factory rather than having all the workers come to your home and build one from scratch in your driveway.

Modular reactors will cost around $400 million and be shippable by truck, rail, or sea to wherever they are needed, such as remote northern towns that currently rely on diesel, island nations, and mining operations. They could also be used to power desalinization, for disaster relief, and as backup power for wind and solar.

SMR reactors will be underground for safety, making the above-ground footprint much smaller than conventional reactors.

Image courtesy of Ultra Safe Nuclear Corporation (USNC)

The design of an SMR is quite straightforward and completely self-contained. Think of a large metallic tube standing on its end. At the bottom is the core with the fuel, which is surrounded by water. Heat from the core turns the water to steam, which rises to the top of the tube where the heat is removed, condensing the steam into water so it flows back down to the bottom in a continuous loop. Above the core are control rods held in position by electromagnets that must be turned on. If there is a power failure, the electromagnets immediately turn off, releasing the rods and causing them to instantly drop down, by gravity, into the core and stop the reaction. Even if the core runs out of water, it will not melt down, so these reactors are much safer than reactors in the past. There are also alternative designs that don't use water at all and are inherently safer. According to Dr. Todd Allen, Glenn F. and Gladys H. Knoll Department Chair of Nuclear Engineering and Radiological Sciences, University of Michigan, Ann Arbor:

A second approach is where the fuel is fundamentally different. It's called a triso-fuel. Instead of the traditional fuel rod, where you have little pellets of fuel in a cylindrical tube, these are graphite spheres, so you put them in the centre and surround it with layers of ceramics. They are designed to be cooled by something like helium rather than water, which allows you to operate at much higher temperatures, but it is almost impossible to melt the fuel.

When people think about the high-consequences, low-probability situation like Fukushima, it becomes very unlikely because of this fuel type. Part of that is because you are putting far less power into a single fuel type.

There are actually two versions. You can take little teeny pellets and roll them into the size of a softball—that's called a pebble bed—or you can take those same pellets and put them into a cylindrical box.

Another version changes the coolant, so we are going to use a liquid metal, like a sodium, or a molten salt as a coolant. The advantage here is that they transfer heat way better than water. This makes it much more difficult to melt the fuel.[7]

Since molten salt has a higher boiling temperature, it will not turn into steam. That means there will be no pressure buildup inside the reactor. The reactor itself will sit in a larger pool of sodium, which will remove even more heat. And there's one other advantage to molten salt reactors that adds another level of safety:

> For the salt systems, most of them use a liquid fuel; they dissolve the fuel in the salt. There are a couple of concepts that just use the salt as a coolant. For the salt coolant ones, there is a different mechanism to prevent accidents. You can drain the salt into different bins, and those bins are too small for the reaction to keep going. That's one approach. You can't damage the fuel because, first, it's already melted, a liquid, and, second, you've got a coolant that transfers the heat away much faster than water. So with the salt design, there are multiple safety mechanisms that go beyond what traditional reactors use.
>
> *Dr. Todd Allen*

Another alternative is called the breeder reactor, which can burn spent nuclear fuel from older reactors and will help with the disposal of nuclear waste. The breeder reactors do have their own waste, but the products are shorter lived.

The sealed modular design of SMRs makes them an all-in-one-package product, which means they are also a commodity that can be sold around the world. A conservative estimate of the global export potential of SMRs is $150 billion.

The idea is to place these modular reactors underground, where the Earth provides natural shielding and where they are out of reach of anyone with evil intentions. Water, or a working fluid, is pumped down where the reactor turns it to steam that returns to the surface to run turbines after which it is recycled back underground as water in a closed loop. The above-ground structure would be a fraction of the size of a conventional nuclear plant, resembling a small factory with no smokestack. Such a system could power about three hundred homes. If the demand for electricity rises, simply add another module.

With a lifetime of twenty years, when the fuel runs out, the modular reactor is simply switched out for a new one, with the old one shipped back to the factory to be reprocessed.

Of course, there are safety concerns with SMRs just as there are with any form of energy production. The biggest threat is flooding from groundwater, which could spread contaminants through the water table. The location of the underground reactor must be chosen carefully to make sure the geological formations are stable enough to minimize risk of earthquakes.

There are also concerns about shipping nuclear reactors around the country by road, rail, and water, but this type of transport is already being done.

> Most people have no idea we are transporting nuclear material in trucks all the time. Even when we talk about nuclear waste, people say there is no solution. I actually think about waste in three bins. There is lower-level waste that comes out of hospitals and nuclear plants, and we dispose of it all the time. In the US, we have defence waste and we have an underground repository; we dispose of things all the time. It's only our commercial spent fuel for which we haven't come up with a solution yet. But when I look at the low-level waste, we are transporting that nuclear waste all the time.
>
> As an engineer, I know we can do that safely. I don't think people know how we do that—how those casks used in transport systems are built, the testing they go through—but that's because we do this out of sight. It's been a reflexive notion from nuclear people: keep your head down, don't be noticed. I think that in the end that just haunts us because it allows people to have nervous opinions about this—nervousness about the technology—without ever having engaged with it.
> Dr. Todd Allen

During my visit to the Bruce Power Generating Station that I mentioned earlier, I asked to see the storage facility for the new fuel rods. I was led to an area containing stacks of cardboard boxes.

"That's it?" I exclaimed. "Yup. It comes in cardboard boxes on a regular transport truck," explained my host. "If you didn't know what it was, you couldn't tell."

My host then opened one of the boxes and handed me a metal fuel bundle. (We were wearing gloves to keep our finger oils off the surface.) CANDU reactors use natural uranium, which is how it comes out of the ground and is not very radioactive; you can literally hold it in your hand. It is only when the uranium is in the reactor and afterwards that it becomes too hot to handle.

"If I had evil intentions, could I turn this into a weapon?" I asked.

With a wry smile, he responded, "About the only way you could hurt someone with that is to club them over the head with it. But you can't make a bomb out of it if that's what you're thinking."

In Canada alone, more than thirty organizations from across the country are opposed to SMRs, including the Canadian Environmental Law Association, Greenpeace, Canadian Coalition for Nuclear Responsibility, and Sierra Club Foundation. The biggest concern among these groups is transportation of nuclear waste. If SMRs are going to burn spent fuel, that is high-level waste that has to be moved from the current storage facilities to reprocessing plants where it can be adapted for the new reactors. One of the elements extracted from the fuel will be plutonium, the element used in nuclear weapons. This poses a danger of nuclear proliferation should enough of that material get into the wrong hands.

The modular reactors are designed to be shipped from the factory with the nuclear fuel already inside. When the reactor comes to the end of its life, the spent fuel will still be inside when it is returned to the factory. That means high-level nuclear materials will be moving by truck, rail, or ship, which opponents say poses a risk to society. Clearly, any transportation of these materials will have to be done in extremely well-controlled circumstances and could be met with public protests.

We need to do better to bring people to the technology and see what it looks like. I think the people who are anti-nuclear are set, and they haven't seen the technology. People ask me: "How are you comfortable with nuclear technology?" And I say:

"Well, I was in the navy, and I had to go on a ship with a nuclear reactor. I know what that means so it's easy for me to understand what's involved in the technology because I've lived on board."

I think we've done a horrible job, keeping people away from the technology. There is an opportunity now, primarily through climate change discussions, to bring people back into this discussion if we do it well.

I know it's going to be hard, but it is really early engagement with the public, which tends not to be something that the nuclear industry does well. You want to have that discussion ten years before you even think about transporting it. I think we are seeing it now on the waste programs. Now that I'm later on in my career, I'm becoming less concerned with technical improvements and more aware of social issues around the technology.

Dr. Todd Allen

Small modular nuclear reactors have the potential to become part of the mosaic of energy sources in the future, providing 24/7 power, first, in smaller communities. Then, as the technology becomes more accepted and SMRs are incorporated into larger urban centres, they can become the steady baseline power that is always there when the alternatives have gone dark and batteries run down.

At this point, they are only designs on paper, with full implementation decades away.

Today's move towards smaller nuclear reactors is comparable to computers thirty years ago, when computing moved from large mainframes to desktop to laptops. The technology is improving, but whether it is adopted depends on public, political, economic, and environmental decisions.

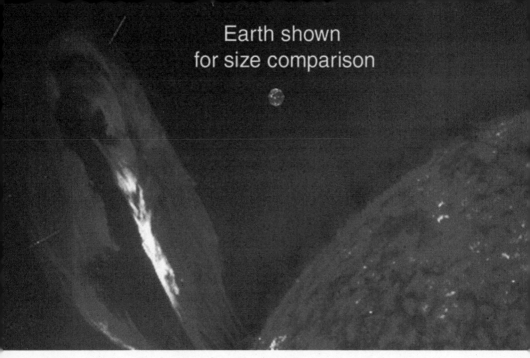

Earth shown
for size comparison

The sun is an enormous fusion reactor. The large eruption
on the left dwarfs the Earth.

© NASA ESA

Chapter Nine

Fusion Power

*We've been promised clean fusion for decades, but it
still has not been achieved. What lies beyond the ITER
Project?*

Industry jokes still making the rounds: "Fusion has been twenty years
away for fifty years" and "Fusion power is just thirty years away and it
always will be."

Look up on a clear night and you will see fusion everywhere. All those
shimmering points of light shining like diamonds in the dark are the
process of fusion producing astounding amounts of energy that enable
stars to cast their light across half the universe. Stars are fusion power
plants that burn hydrogen, the most common element known, and
they run for billions of years. The closest star to us, our sun, has been

166

driving just about every form of energy on the surface of the Earth for four and a half billion years and is only halfway through its life. So, you could say that fusion is the cheapest, most efficient, reliable, and longest-lasting energy source anywhere.

Reproducing that process on Earth holds the promise of abundant, clean energy for the future, but after more than half a century of research, that has yet to be achieved. Ironically, a process that is so common throughout the universe is extremely difficult to replicate here on Earth.

It is relatively easy for stars to achieve fusion simply because they are so big. Our sun, an average-sized star, is one hundred and nine times wider than the Earth. To scale the comparison down: if the sun was the size of a basketball, the Earth would be not much bigger than the head of a pin. About a million Earths could fit inside the sun. That's inconceivably big. And even though the sun is mostly made of lightweight hydrogen gas, there is a lot of it, so the sun's mass is huge, more than three hundred thousand times that of the Earth. All that mass produces powerful gravity, so if you could stand on the surface of the sun (not recommended), you would have a difficult time holding yourself up because you would weigh twenty-eight times your weight on Earth. An average person would tip the scales at about 2 tons.

All of that gravity squeezes the sun in towards its centre and puts the core under such tremendous pressure that it heats it up to 15 million°C, which provides the energy to not only fuse together atoms of light hydrogen to make heavier helium but also release massive amounts of energy in the process. The reaction is self-sustaining if the pressure and temperature remain high and there is a supply of fuel. Fortunately, the sun has such a huge amount of hydrogen that it won't run out for a few billion years.

Fusion is the opposite of fission. Instead of breaking atoms apart, as we do in conventional nuclear reactors, fusion forces atoms to melt together, like joining two water drops on a table to make a heavier blob or smacking two handfuls of snow together to make a snowball. Except the new product is not quite the sum of its parts. There is stuff left over, products of the reaction that get cast off at tremendous speed carrying kinetic energy. That extra energy from a fusion reaction, which is several times greater than the energy from fission, can be captured to generate electricity.

Fusion power is also very clean, with none of the nasty radioactive by-products that come out of conventional reactors and can last thousands of years. It holds the promise of abundant clean energy. At least that is the hope.

To fuel a fusion reaction, you need two forms, or isotopes, of hydrogen: deuterium, which can be easily obtained from water so there is no shortage of supply, and tritium, which can be made from lithium, the stuff found in batteries. And you don't need very much of either.

According to Dr. Stephanie Diem, assistant professor of engineering physics at the University of Wisconsin–Madison:

> If you extract the deuterium from one bathtub full of water, use the lithium from one battery to make tritium, and combine the two, that will produce enough energy for the lifetime of one human. It could drive your car, heat your house—all the needs of one human. If you compare that to burning coal, it would take 230 tons of coal for the equivalent amount of energy. And that coal puts out a heck of a lot of pollution.[1]

Deuterium and tritium fuse together at a slightly lower temperature than pure hydrogen, which makes the task a little easier. Still, that's a lot of energy packed into very little fuel, and the only by-product is helium, which is a harmless gas that's great for party balloons.

There is just one tiny little hurdle to overcome: before you can get that fusion energy out, you have to pump a tremendous amount of heat into the reactor to make fusion happen in the first place. Atoms don't easily fuse together. The forces that hold individual atoms together are so strong that they oppose the forces of other atoms. Try pressing your thumb through the top of a table. You will feel resistance where the table seems to be pushing back. Even though the atoms in your hand and those in the table are 99 percent empty space, the nuclear forces that hold atoms together are so strong they don't easily break. What you are really feeling is the atoms in your thumb resisting the atoms in the table and vice versa. If you could press down with millions of tonnes of pressure at millions of degrees of heat, your thumb and table would fuse together. Of course, that would be your last act because the fusion energy released in the process would vaporize both you and the table.

Here on Earth, it is extremely difficult to produce the unbelievable pressures that exist at the core of the sun. So, instead of using gravitational pressure, fusion reactors use tremendous heat to give the atoms enough energy so they will fuse together. The heat required is between 100 and 200 *million*°C, which is more than ten times the temperature at the centre of the sun. The only place in the universe where temperatures are that high is inside supernova explosions, where giant stars blow themselves to bits and produce enough light to outshine a galaxy. If the sun went supernova (which it won't; it's not large enough), the Earth would be incinerated. That is the kind of temperature we want to contain in a fusion reactor. There are no materials on Earth that can survive those temperatures without melting or being vaporized into a gas. In blast furnaces, temperatures reach thousands of degrees, but how do we contain millions of degrees of heat? So, the first big challenge to sustaining fusion reactions is to keep the hot fuel from melting the container that holds it.

The second problem, related to reaching that insane temperature, is that it takes a huge amount of energy, and so far, there is more energy being pumped into these devices than comes out from fusion. No one has been able to reach ignition, a break-even point, where the fusion reaction sustains itself so that the energy coming out is greater. Even though fusion reactions have been achieved in experiments, they have only lasted for seconds at a time. The goal is to produce what the researchers call a "burning plasma" that will continue like a super-hot candle flame that burns on its own after it is lit.

One approach to achieving fusion is using a device called a tokamak. The name comes from a Russian acronym that translates to "toroidal chamber with magnetic coils." A toroidal shape, or torus, is basically a doughnut. A fusion reactor looks like a hollow metal doughnut large enough to stand up in and walk around inside. Surrounding the device are superconducting magnets, which are large coils of wire cooled to hundreds of degrees below zero Celsius.

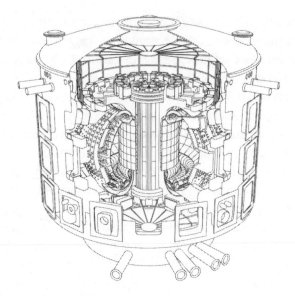

Cross-section of a tokamak reactor.

Image courtesy of ITER

At that low temperature, electrical resistance virtually disappears in the wire, so huge currents can be run through without losses and without heat. Those currents produce powerful magnetic fields that are the same doughnut shape as the inside of the chamber.

A Brief History of Superconductors

Superconductors were discovered in 1986 by IBM researchers J. Georg Bednorz and K. Alex Müller, who went on to win the 1987 Nobel Prize in Physics, the shortest time between discovery and prize. They used a mix of exotic materials: lanthanum, yttrium, and barium copper oxides. The acronym ReBCo is often used to describe them. When an electric current was applied at very low temperatures, the current would flow through as though the material wasn't even there, unlike regular copper wires that absorb some of the electricity and turn it into heat. These first superconductors were in the form of black crystals, which were difficult to work with on a larger scale. Decades of research have resulted in thin strips that can be laid on stronger steel or copper that can be rolled out for superconducting magnets.

The idea in a fusion reactor is to first create a super-hot plasma, which is an ionized gas that has an electric charge. Neon signs use ionized plasmas that glow red inside the glass tubes. Lightning bolts are white-hot plasmas created when static electricity in clouds heats up and breaks down air. There are two gases in the fusion reactor. Deuterium, also known as heavy hydrogen, has a nucleus containing one proton and one neutron. A regular hydrogen nucleus is just a single proton. Deuterium occurs naturally in oceans so it is very abundant. The second gas, tritium, has one proton and two neutrons. Tritium can be obtained from neutrons interacting with the metal lithium and is slightly radioactive, often used as a tracer for medical purposes or in self-illuminating toys. These heavier atoms are used because they will fuse more easily than pure hydrogen and will do so at a lower temperature (if you call above 100 million degrees low!).

To start a reactor, a small amount of the neutral gases are injected into the chamber. Powerful magnets are then turned on to strip electrons off the atoms, which turns the gas mixture into an electrically charged plasma that glows like a neon sign but much brighter. According to Dr. Diem:

> We drive a current through the central solenoid magnet, which induces a current in the plasma. And just like when you drive a current through a copper wire, the plasma heats up. It also ionizes the rest of the gas that is in the vessel. It is very much like a transformer, where you run a current in the primary winding and it induces a current in the secondary. In this case, the solenoid in the centre is the primary winding and the plasma is the secondary.[2]

It takes a lot of magnetic coils placed strategically around the device to make what is referred to as a magnetic bottle to contain the plasma and to compress it into a ring that is suspended in the middle of the hollow doughnut, not touching the walls or any surface. That is the big secret to containing the high temperatures without melting the container: don't let the hot plasma touch anything by levitating it in mid-air.

Forming this magnetic bottle has been one of the big challenges in fusion research up until recently. The magnetic field formed by the superconducting coils has to be perfectly shaped so that the plasma forms a smooth ring. Meanwhile that plasma is trying to escape its magnetic confines. The task of controlling the plasma has been compared to trying to contain jelly with a rubber band. Any distortions in the magnetic field will cause turbulence in the plasma and cool it down, stopping the reaction. The magnetic bottle must be strong enough to keep the plasma ring together in a smooth shape. The next step is to make that plasma super-hot.

There are a few different ways that plasma can be heated to super-hot temperatures, with the two main techniques being injecting neutral particles, also called neutral beam injection, and injecting radio frequency waves. In the first method, a beam of highly energetic fast-moving neutral particles can be shot into the device. As they hit the plasma, the neutral particles are ionized and give up some of their energy to heat the rest of the plasma. The other method injects radio or microwaves.
Dr. Stephanie Diem

As the free electrons in the gas follow the magnetic field around the doughnut, they move through corkscrew spiral paths that give them an up and down motion. The microwaves are tuned to that up and down frequency, pumping more energy into the system, the way pushing a swing in a park at the right frequency makes the swing go higher. In the plasma, this adds more heat.

The three processes—microwaves, magnetic induction, and neutral beam injection—combine to drive the temperature higher and higher until somewhere between 100 and 200 million°C, deuterium and tritium nucleii begin fusing together to form helium 4, with two protons and two neutrons. This is the most stable form of helium and the second-most-abundant substance in the universe after hydrogen.

Now you would think that once you brought something up to a 100 million°C, it would be fairly easy to tap into that heat to make electricity. But if you tapped directly into the plasma, you would be losing energy. That heat is needed to keep the fusion reaction going.

It takes a huge amount of electricity to run those big magnets and make the plasma in the first place, so if you remove any heat, you are taking away some of what you put in, which is a loss. You would be better off just using the electricity you started with.

The beauty of a fusion reaction is that once it achieves ignition, it gives more energy back than we can use. Not heat, but rather a flow of neutrons that are left over once fusion takes place. It goes like this.

Nuclear Fusion

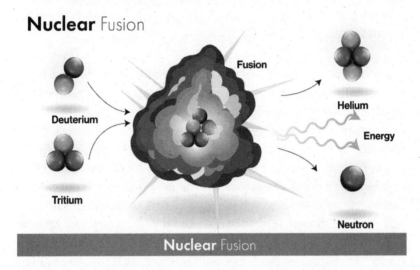

A fusion reaction fusing deuterium and tritium into helium and an extra neutron that is given off.

© uday/Adobe Stock

Deuterium has one proton and one neutron. Tritium has one proton and two neutrons. So when they fuse to become helium 4 with two neutrons and two protons, there is a neutron left over. This is a case where the new product is less than the sum of the original two parts. That difference in mass is converted to a lot of kinetic energy, the energy of motion, so the extra neutron is ejected from the plasma at very high speed. Because it is neutral with no electric charge, it is not contained by the magnetic field: it escapes, carrying a lot of energy with it. As long as the fusion reaction can be sustained, billions of escaping neutrons will carry more energy out than was put in.

Now we're talking.

It might seem counterintuitive to get more out of a system than you put into it, but fusion is a testament to the powerful forces contained within the nucleus of an atom. And we can thank Professor Einstein for telling us just how much energy we get when those forces are released.

Surrounding the reactor are blanket materials that stop those speeding neutrons and turn their kinetic energy into heat, which is used to boil water, make steam, and run a turbine to generate electricity. PHEW!

That's the concept. Getting everything to work perfectly for a long period of time is the challenge.

Many of these stages of a fusion reaction have been achieved separately in different devices around the world, but not together for a sustained period in one device.

Tokamaks were invented in Russia in the 1960s and several countries have built them for fusion research. The largest operating fusion reactor is the Joint European Torus (JET) in England, which at 16 megawatts holds the record for the most energy produced by a fusion reaction. Another, the Wendelstein 7-X in Germany, did a test run reaching almost 180 million°C, and the Chinese EAST Reactor sustained a fusion reaction for 102 seconds.[3] However, none of these devices have reached the break-even point. That is the job of ITER.

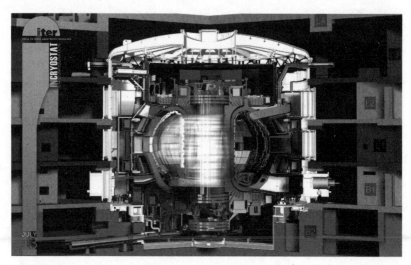

ITER fusion reactor.   Image Courtesy of ITER

In southern France, the world's largest fusion reactor, called ITER, is under construction and will be twice the size of the one in England. ITER means "The Way" in Latin, and this international collaboration by thirty-five countries will be the first to produce a sustained fusion reaction that is expected to put out ten times more energy than required to run it. Even if it only puts out twice the power, the scientists will be happy to have achieved the major milestone of passing the break-even point. Once sustained fusion is achieved, the goal is to use 80 percent of the energy to make electricity and pump 20 percent back into the reactor so it becomes self-sufficient. In that way, the plant runs completely on its own, with no reliance on outside power. It will be a little piece of sun shining here on Earth.

By the way, if you are worried about a device designed to replicate the nuclear explosions taking place at the centre of the sun exploding—it won't.

We use power supplies to energize the magnets that contain the plasma and produce the fusion, but as soon as we cut off the power supply, the fusion reaction just goes away instantaneously. The particles then recombine and form a neutral gas again. Like turning off a light, it just goes out.
Dr. Stephanie Diem

ITER is designed to produce 500 megawatts of power from an input of only 50 megawatts. It is a research reactor meant to demonstrate the technology that will create the ultra-hot conditions for fusion to take place and to provide the foundation for future commercial reactors.

We've done a lot of advanced work in how to contain a magnetic field, reached record temperatures, and confined it for a long time. We have made amazing advances in computing technology right now, which basically allows us to model the whole device, which is an amazing tool when you go to design something at a reactor scale. We have advances in additive manufacturing, so you can produce prototypes and advanced alloys that can withstand these environments.

And then we will have access to ITER, which will be the first time we have a burning plasma. Advances in high-temperature superconducting magnets may allow us to construct more compact reactors. With all these things, which we now have for the first time, we can really realize commercial energy.
Dr. Stephanie Diem

The agreement to build ITER goes back to 2006 when the European Union, China, India, Japan, Korea, Russia, and the United States joined together not only to provide funding for the $15-billion project but also to build component parts in the individual countries that would then be brought together and assembled on-site. So why has it taken so long to build one reactor?

The project was supposed to be completed by 2016, but the international co-ordination of manufacturing elements in different countries has been one of the big causes of delays that have stretched the project out over thirty-five years and ballooned a budget that is expected to top out somewhere between $30 billion and $60 billion. But construction is underway (70 percent complete in 2021) with the production of the first plasma expected by 2025 and full fusion in operation by 2035.[4]

Once ITER is up and running, its prime role will be to study the integration of supporting systems to make fusion work, including materials that make up the chamber and the blanket materials that absorb the heat. Both are degraded by the bombardment of neutrons, so work needs to be done to develop more durable materials that can withstand a long-term fusion reaction.

In the meantime, the Massachusetts Institute of Technology announced they have a smaller, cheaper design that will be able to achieve fusion in the same time frame at less cost. Called SPARC, it will use the same principles as ITER but take advantage of new developments in superconducting magnets, enabling it to achieve the same plasma temperatures in a much smaller device. When completed, SPARC is expected to deliver twice as much energy as is pumped into it. This will become a prototype for future commercial fusion reactors, producing about 150 megawatts of electrical energy.[5]

The SPARC fusion reactor is much smaller and more efficient than ITER due to advancements in superconducting magnets.

Created by Ken Filar, courtesy of PSFC

The new superconducting magnets in SPARC operate at a relatively warm temperature, just above -200°C, which will still take your fingers off, but it is warmer than temperatures used in ITER, which need to be cooled down with liquid helium to -269°C. That may not seem like much of a difference, but it means the new superconductors can be cooled with much cheaper liquid nitrogen rather than liquid helium and can produce more powerful magnetic fields in a smaller space. SPARC will be about half the size of ITER. Unfortunately, the newest versions of these superconducting magnets were not available when ITER was designed, and that project is too far along to be retrofitted.

A report in *National Academies of Sciences, Engineering, and Medicine*, "Bringing Fusion to the US Grid (2021)," says the challenge is designing a concept by 2028 at a price point that commercial industry will buy into. Then it will be a private-public partnership from 2035 to 2040 with power plants designed for 50- to 100-megawatt electric and a price point of $5 billion to $6 billion for the initial demonstration.[6] Hopefully, future power plants will be cheaper.

Depending on market needs, this size of projected fusion reactors could potentially be similar to the small modular nuclear reactors intended for regional use rather than the large gigawatt power plants of the past. The advantage of fusion will be that it is clean energy with no radioactive by-products and that it will provide baseline power for the grid that is available 24/7. The challenge will be finding investors willing to cover the high upfront costs of construction.

Now, if gigantic magnets and plasma toruses are not to your taste, how about a more Star Wars–style approach to fusion?

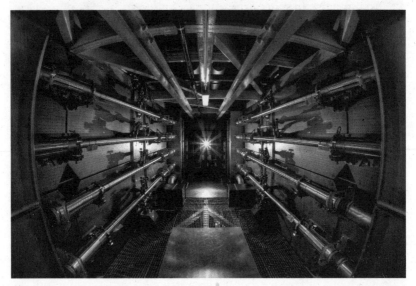

National Ignition Facility at Lawrence Livermore uses a bank of powerful lasers aimed at a fuel pellet the size of a peppercorn.

© Lawrence Livermore National Laboratorycom

Imagine 192 giant lasers, each capable of slicing through your body like a light sabre, pointed at a little ball the size of a peppercorn. It's like the spy who is discovered in the villain's lair and suddenly faces guns from all directions. This alternate approach to fusion uses the energy from lasers to blast atoms together, not unlike the way the sun does it. The lasers fire all at once, squeezing the ball down with such force that the hydrogen fuel inside fuses and energy is released. Remember, according to Einstein's $E = MC^2$ formula, a small amount of mass can release a lot of energy.

> You have a pellet, and when a laser interacts with a surface, that material begins to ablate away, or essentially move very quickly in the opposite direction. If you have a sphere, you have the fuel capsule blowing out in all directions. And just like with

the rocket effect—every action has an equal and opposite reaction—you've got a very strong force moving outwards, so you're going to get a very strong force moving inwards.

That starts the compression process. You want to do that fast enough and symmetric enough so that the deuterium and tritium fuse and create thermonuclear burn.

Dr. Carolyn Kuranz, associate professor of nuclear engineering and radiological sciences, University of Michigan

You would think this technique, akin to shooting fish in a barrel —or more like shooting a flea with a group of cannons—would be fairly straightforward, but it has turned out to be far more challenging than anyone imagined. For more than sixty years, scientists at the Laboratory for Laser Energetics at the University of Rochester have been developing lasers powerful enough to get tiny pellets about the size of a pinhead to fuse. In 2009, the National Ignition Facility in California joined in the research, using more powerful lasers that put out 2.15 megajoules, which is the equivalent of millions of flashbulbs going off in a billionth of a second.[7] The target is very small and the laser bombardment very powerful so you might think it would be easy to trigger a fusion reaction, but it has proven to be an enormous challenge.

One approach is to suspend the fuel pellet in a container called a hohlraum, in this case, a cylindrical gold container the size of a pencil eraser.

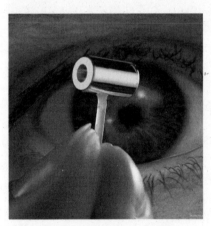

A gold-coated fuel pellet container known as a hohlraum.

Courtesy of Lawrence Livermore National Laboratory

It is essentially a soda can of gold with the pellet in the centre. Instead of the lasers directly hitting the pellet, they're hitting the gold which gets hot. So they are surrounding the pellet in a thermal radiation bath. But then from there, it ablates the

material, blasting it outwards, and that moving out causes compression moving inward that compresses the fuel. It is a different kind of mechanism, but the physics is all the same.
Dr. Carolyn Kuranz

Extremely powerful pulses of laser light are directed at all sides of the target at once, blasting off its outer coating and creating shock waves that are focused towards the centre, causing an instant implosion. The speed of that implosion is incredibly fast, 400 kilometres per second. At that speed you could circle the Earth in one minute. The sudden collapse produces tremendous pressure, 350 billion times greater than the Earth's atmospheric pressure, and temperatures of over 100 million°C, which compresses the pellet down to something many times denser than lead. Theoretically all of that forces the deuterium and tritium fuel to fuse.

Despite decades of effort, the researchers have yet to achieve the break-even point, or get more energy out than is put in. Even though the target is very small—atomic level—the shock waves must all arrive at the centre at exactly the same time. Any imperfections in the outer shell will cause some parts of the fuel to ignite too early, so the pressures and temperatures are not high enough. The lasers must be perfectly synchronized so the beams arrive on all sides of the pellet at precisely the same moment and with the same intensity.

Forming a symmetric shape, instead of a pancake or a sausage, is a challenge, as is getting the shock velocity high enough that it compresses the fuel quickly. You have a shock wave, which will always get to the centre and stagnate, but if it doesn't happen quickly enough, the fuel will not ignite.
Dr. Carolyn Kuranz

In August 2021, the National Ignition Facility managed to spark a fusion reaction that returned 70 percent of the energy put into it. While not break-even, it was a major breakthrough; they had been working for years and getting only 3 percent back. As with other fusion researchers around the world, the promise of clean abundant energy is a powerful motive to keep going.

When fusion-powered electric generators do appear—in another twenty or thirty years—they will likely be the magnetic confinement type, such as ITER, rather than the laser type. To produce sufficient energy out of the tiny pellets, there would have to be a rapid series of fusion implosions. After each firing of the lasers and subsequent fusion reaction, a new pellet and its hohlraum would have to be placed in precisely the right location at the exact centre of the device so all the lasers will strike it symmetrically—and that same process is repeated over and over every few seconds. The technology to accomplish that does not yet exist, nor may it be practical.

This is not to say that laser fusion is a waste of time. On the contrary. The research has produced advances in laser technology, including the 2018 Nobel Prize in Physics awarded to Canadian scientist Dr. Donna Strickland, who worked at the Laboratory for Laser Energetics for her development of ultra-short laser pulses that can be used in medicine, industry, and communication. At the same time, the ultra-high temperatures and pressures produced in the fuel pellets provide insights into ultra-dense materials, such as those found in exotic objects like white dwarfs and neutron stars in space. And of course, the military has always had an interest in powerful lasers.

While most fusion research struggles with the issue of huge amounts of energy required to run these devices, Canadian company General Fusion is taking a clever alternative approach that does away with giant superconducting magnets and super-powerful lasers. It hopes to duplicate the sun's incredible gravitational force and achieve fusion by squeezing.

Diesel engines work the same way, where fuel and air are compressed in a cylinder until it gets hot enough to ignite without the use of spark plugs.

In the General Fusion reactor, a series of powerful pistons surround a container of liquid metal with the hydrogen plasma in the centre. The pistons mechanically squeeze the liquid on all sides at once, heating the fuel by compression. Theoretically, this technique, called magnetized target fusion, can reach the 150-million-degree-Celsius mark in a small space. That heat is absorbed by the liquid metal and used to produce steam that runs a turbine to generate electricity.

The Canadian General Fusion reactor uses extreme pressure to generate ultra high temperatures.

Image courtesy of General Fusion

The General Fusion project holds enough promise that the United Kingdom Atomic Energy Agency has invited the firm to build a prototype demonstrator plant in England to show how the technology works on a commercial scale. It is expected to be operational by 2025 at an estimated cost in the $400-million range, which might seem like a lot but is way below the $20-billion ITER Project.[8]

The promise of fusion energy might be closer than it has ever been before. ITER is nearing completion. Other fusion reactor designs are in development, such as the smaller SPARC at MIT and one called a stellarator, which could require less input energy than a tokamak and which is being tested in the United States, Germany, and Japan. And who knows, perhaps the Canadian approach will be the first to turn that long promise into a reality?

While the world continues to wait for fusion power, the cost of clean alternatives continues to drop. If commercial fusion plants end up being smaller units producing 50 to 100 megawatts of electricity at a cost of a few billion dollars, the same amount of electricity could be generated by ten large wind turbines at a cost of about $10 million each and they could be installed in a very short time. It may come to a point where the alternatives are so cheap that fusion energy will occupy niche markets, such as providing baseline power, powering remote communities, or supporting military applications. Whatever the case, let's hope that fusion does not become one of those bright ideas forever destined to remain in the future.

Global fossil fuel consumption

Global primary energy consumption by fossil fuel source, measured in terawatt-hours (TWh).

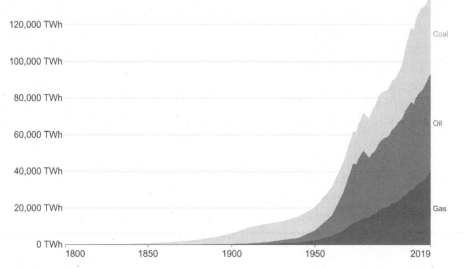

Source: Vaclav Smil (2017). Energy Transitions: Global and National Perspective & BP Statistical Review of World Energy
OurWorldInData.org/fossil-fuels/ • CC BY

Graph showing world oil consumption.

Chapter Ten

Rethinking Oil

Can we extract that energy with techniques other than straight combustion?

Today we face a conundrum. Climate change is providing the incentive to move away from oil, an incredibly convenient, energy-rich resource. But to abandon the golden egg entirely overnight would have disastrous economic impacts since so many livelihoods depend not only on drilling, transporting, refining, and distributing oil products but also on building and maintaining the myriad machines that run on oil or other fossil fuels. Indeed, modern civilization, since the birth of the Industrial Revolution, was built on coal, oil, and natural gas. Remove that foundation all at once and civilization as we know it would crumble.

There are still about 1.3 trillion barrels of oil reserves worldwide, which, at the current rate of consumption, is enough to last about fifty years. At the same time, carbon emissions are continuing to rise, warming the climate, so we have a race between those who want to end our reliance on oil and those who depend on it.

To get an idea of how ingrained fossil fuels are in our modern lives, look around you right now. It goes far beyond the gas and oil you put into a vehicle. You are surrounded by the products of fossil fuels. Even if you don't own a car, the petrochemical industry has supplied you with plastics used in your computer case, the soles of your shoes, urethane coatings over wood floors, plastic food containers in your refrigerator, kitchen utensils, toys for kids, and more. Then there are all the cleaning products, the synthetic fibres in the clothes you are wearing. Depending on where you live, the lights in your room could be fossil powered. And that's not the half of it.

Most fossil fuels are consumed to provide transportation, heating, and electricity; to make fuels such as gasoline and diesel; and, through large industries, to provide the energy needed to make the steel, aluminum, and glass in everything from vehicles to highrise buildings. Even if you try to do your part to curb climate change by walking instead of driving, the concrete sidewalk under your feet needed fossil fuel for its construction and more CO_2 was released by the chemical processes in the manufacture of the concrete itself.

There used to be a worry that we would run out of oil—after all, it is a limited resource. But now, with oil wells drilled deeper into the Earth; sources found farther offshore, or in more remote areas such as the Arctic; and the newer technology of fracking, which has unleashed vast reserves of natural gas, supply is no longer the major concern. The biggest issue is what to do with what is left.

According to Our World in Data, a research group out of Oxford University:

Concerns surrounding this risk have persisted for decades. Arguably the most well-known example of this was Hubbert's Peak Theory—also known as the Hubbert curve.

M. King Hubbert, in 1956, published his hypothesis that for any given region, a fossil fuel production curve would follow a bell-shaped curve, with production first increasing following discovery of new resources and improved extraction methods, peaking, then ultimately declining as resources became depleted.

During the 1979 oil crisis, Hubbert himself incorrectly predicted the world would reach "peak oil" around the year 2000; and in the decades since, this prediction has been followed by a succession of premature forecasts by analysts.

Meanwhile, actual global oil production and consumption continues to rise.

If we look at trends in proven fuel reserves, we see that our reported oil reserves have not decreased but *increased* by more than 50 percent, and natural gas by more than 55 percent, since 1995.

Based on BP's Statistical Review of World Energy 2016, we'd have about 115 years of coal production, and roughly 50 years of both oil and natural gas remaining.

In the latest Intergovernmental Panel on Climate Change (IPCC) report, the budget for having a 50 percent chance of keeping average warming below two°C was estimated to be approximately 275 billion tonnes of carbon.

Here's the crucial factor: if the world burned all of its currently known reserves (without the use of carbon capture and storage technology), we would emit a total of nearly 750 billion tonnes of carbon. This means that we have to leave around two-thirds of known reserves in the ground if we want to meet our global climate targets.

So whilst many worry about the possibility of fossil fuels running out, it is instead expected that we will have to leave between 65 to 80 percent of current known reserves untouched if we are to stand a chance of keeping average global temperature rise below our two-degrees global target.[1]

If most of the remaining fossil fuels need to remain in the ground for reasons of climate change, but those reserves support a large portion of the economy, is there a way to still extract value from those fuels without heating up the atmosphere?

There may be a way out of this conundrum if we rethink how energy is extracted from fossil fuels.

Oil, coal, and natural gas are not really the bad guys in the climate crisis. It is the way we have been burning them that is warming the atmosphere and acidifying the oceans. For more than a century we have basically been digging these fuels out of the ground and setting a match to them. Consider coal. It is a rock that burns. Hold a flame to the black rock and it will burn entirely on its own. The Chinese discovered this about four thousand years ago, using coal for heating and cooking.[2] It was also adopted in medieval Europe as the "black stone," although it was considered inferior to wood because of the smoke and soot. However, when wood became scarce, coal took over as a European household mainstay.

The ancient Chinese also discovered petroleum, around 2000 BCE, from oil seeps in the northwest part of the country. It was used in lamps, as lubricants, and even in medicines. Around 200 BCE, the Chinese were the first to use natural gas. To extract salt from brines, evaporators were heated by a flammable gas conveyed from the ground through bamboo pipes.[3]

The biggest move beyond heating and cooking with coal, and what many consider the beginning of the Industrial Revolution, was Englishman Thomas Newcomen's invention of the atmospheric engine in 1712. The device, which was the size of a house, was the first steam engine and was used to pump water out of mines. It was so successful that the design became popular throughout England,

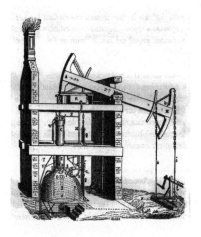

Cross-section of a Newcomen atmospheric steam engine.

© Engineering and Technology History Wiki

allowing miners to dig deeper into the Earth than ever before without fear of flooding the mine tunnels.

Steam engines are simple in concept: boil water using whatever heat source is available and take advantage of the fact that when it reaches 100°C and turns to steam, water expands by seventeen hundred times. Most steam engines use that pressure to drive a piston in a cylinder, but surprisingly, Newcomen's device used the opposite effect: that of cooling steam condensing back to water and contracting by seventeen hundred times. In his machine, steam was introduced underneath the piston when it was at the top of the cylinder, then cold water was sprayed in to make the steam condense and create a vacuum. The atmospheric pressure from the outside is what drove the piston down (through the vacuum) and provided the power.

How Air Pressure Works

You can see the power of atmospheric pressure using an empty soda can, a heat source, a pair of metal tongs, and a pot of cold water.

What to do:

1. Add a small amount of water to the soda can, just enough to cover the bottom.

2. Place the can on the burner of a stove and turn the heat to high.

3. Place the pot of cold water close to the can on the stove.

4. When steam comes out the opening of the can, use the tongs to pick up the can and quickly plunge it upside down into the cold water.

The soda can will instantly implode when the steam inside condenses and creates a vacuum. In less than a second, atmospheric pressure on the outside crushes the can into a crumpled mass with a loud pop. It is a dramatic demonstration of the power of air pressure.

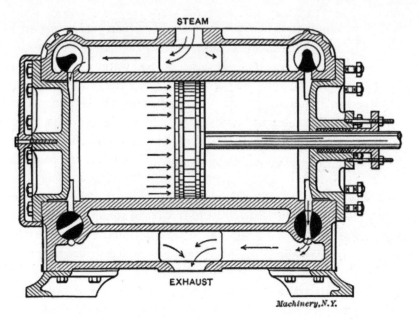

Cutaway of a steam engine. The piston is driven on both directions by the steam.

Image courtesy of The Gutenberg Project.BM

Later, improvements on the design of steam engines used the pressure of expanding steam to drive a piston through a cylinder directly. When the piston reaches the end of its travel, the steam is directed to the other side, pushing the piston back: it is driven in both directions, providing twice the power. The back-and-forth motion of the piston is converted into the circular motion of a wheel, and you have steam power. It doesn't matter how you heat the water, just keep it boiling and the piston keeps moving, providing more power than muscle, beasts, or wind.

Steam engines drove the first part of the Industrial Revolution, giving power to factories and, eventually, to rail transportation through steam locomotives, with coal as the fuel of choice because it burns hotter and was more abundant than wood. The coal was just shovelled into boilers and burned directly to keep the water boiling, resulting in a forest of smokestacks towering over England, spewing clouds of soot, blackening buildings in London, and contributing to the city's famous fog.

Otto 4 cycle internal combustion engine. It resembles a steam engine but the combustion happens inside the cylinder so there is no external boiler. This same principle is used in modern combustion engines.

Image courtesy of Vintage Machinery

More than a century later, in 1864, to improve energy production, pioneering engineers Nikolaus August Otto and Karl Benz modified the steam engine. Rather than burning the fuel on the outside to boil water, they brought the combustion inside the cylinder to drive the piston directly by gas explosions. The more efficient and cleaner-running internal combustion engine, the 4-cycle from Otto took over from steam.[4] They were first used as stationary industrial engines in factories, then Karl Benz improved on the design, made it smaller so it became the power plant of choice in automobiles right up to the present time.

Of course, these new engines could not run on coal because it is very difficult to get powdered rock in and out of a cylinder at high speed without grinding the machinery to bits. The engines first employed a gas by-product of coal, sometimes known as Town gas, that was used in gas lamps in London homes and streets. Made by heating coal without oxygen, it was a mixture of hydrogen, methane, and carbon monoxide. The gas could be easily brought into a cylinder, burned, blown out as exhaust, and replaced with new gas. Later, the coal gas was superceded by distillates of oil such as gasoline, diesel fuel, and alcohol, spurring a whole new demand for oil.

As transformative as fossil fuels have been to modern civilization, for the first hundred years or so, no one cared about the exhaust gases and soot produced by these fuels. Tailpipes and smokestacks just discharged them into the air where they simply dispersed. Out of sight, out of mind. But as the great smog of London in 1952 and, later, smog over Los Angeles in the 1960s and '70s showed, those gases can accumulate in the air affecting health—and now, altering the global climate.

A number of laws have been introduced over the decades to minimize the impact of exhaust gases, such as the Clean Air Act in the United States that demands catalytic converters on vehicles and scrubbers on smokestacks to remove pollutants such as sulphur dioxide and nitrous oxide. However, those measures have not been enough to curb the release of carbon dioxide, which is heating up the atmosphere.

Part of the reason for slow progress on cleaning up the air has been pushback from industries that feel threatened and have lobbied some governments to roll back those regulations in an attempt to keep business as usual. But that doesn't mean governments and the public are unable to make changes when necessary, as all of North America discovered during the oil embargo in the early 1970s.

During the Arab-Israeli war of 1973, members of the Organization of the Petroleum Exporting Countries (OPEC) imposed an embargo against the United States for their military support of Israel. Since the US was dependent on Middle Eastern oil, fuel supplies across the country simply dried up, and to the shock and horror of people everywhere, gasoline stations simply ran out of gas. Overnight, the price of oil and, therefore, gasoline shot through the roof. This came as a complete shock: the North American car culture was used to filling up at gas stations that were on just about every corner. Suddenly, those stations were out of fuel.

The scarcity of oil and gas, which drove prices skyrocketing, was a hard hit in the pocketbook for everyone in North America. Suddenly, attitudes towards large vehicles with huge engines, "gas guzzlers," changed overnight. Smaller, more fuel-efficient Japanese and European vehicles became the rage, while the North American manufacturers with their giant cars scrambled to develop smaller cars to keep up. Where horsepower was once the big selling point, now it was fuel efficiency.

Following the oil embargo, governments poured funds into research on alternative energy—wind, solar, hydrogen, and geothermal—as well as promoting efficiency of buildings through better insulation. It was a positive move towards a clean energy future. But since then, following negotiations with the Middle East and the opening up of more oil and gas reserves in North America, the price of gasoline has remained relatively low, allowing the continued growth of both the oil and automotive industries using the old technologies. So we find ourselves once again with large vehicles driven by large internal combustion engines and even a revival of the muscle car era with sports cars producing up to 700 horsepower and burning gas like there is no tomorrow.

The urgency to get away from the dirty old technology is not as visible today as it was in 1973. Climate change is not as obvious as a sign at a gas station saying, "Out of Gas."

We are entering a new phase of the Industrial Revolution, similar to when combustion engines took over from the steam era. The well-worn internal combustion engine, which is now more than a century and a half old and has been tweaked and refined about as far as it can go, is being replaced by a cleaner, more-efficient electric drive that does a better job.

However, that doesn't mean fossil fuels need to be abandoned entirely. Scientists are looking at ways to extract that energy and use it in new clean technology without producing those carbon emissions. And incredibly, some of these techniques involve tapping into the oil while it is still in the ground.

In simplest terms, all fossil fuels are chains of carbon atoms, with hydrogen atoms stuck along the chain like ornamental lights on a string, hence the name, hydrocarbons. The simplest is natural gas, or methane, which is one carbon atom and four hydrogen atoms. Propane, used in backyard barbeques, is four carbon atoms with ten hydrogen atoms. The heavier the fuel, the longer the chain, so gasoline is eight carbon and eighteen hydrogen; diesel fuel is twelve carbon and twenty-six hydrogen; all the way up to tar, asphalt, and bitumen, which can have forty to seventy carbon and more than a hundred hydrogen.

When we burn the hydrocarbons, it is the hydrogen that comes off and combines with oxygen to produce water (H_2O), releasing energy along the way. You could say that vehicles running on gasoline are actually hydrogen powered. The problem is what happens with the rest of the hydrocarbon that is left behind and released into the atmosphere. Some of the carbon combines with oxygen to make carbon dioxide (CO_2) and a little of the carbon is left behind to form black soot. There are other chemical combinations that can take place, such as sulphur impurities that can combine with oxygen to make sulphur dioxide, which when mixed with water in the atmosphere becomes sulphuric acid, or acid rain.

If burning gas is essentially burning hydrogen, what if we were able to use just the hydrogen and leave the rest of the carbon in the ground? Hydrogen is totally clean because when it burns as a fuel, or is run through a fuel cell, it combines with oxygen to make H_2O or water. That's all that comes out of the tailpipe, and it can be easily recovered and used for other purposes. There are always uses for water.

It turns out this can be done, but it comes at a cost. It takes energy to remove the hydrogen from fossil fuels, so we can't think of hydrogen as an energy source. It's just a cleaner fuel that has to be manufactured.

Switching from an oil economy to a hydrogen economy, as we will see in the next chapter, would not require a huge change of infrastructure. Hydrogen can be transported as a compressed gas or liquid (although in liquid form it is super cold at -252°C, which presents other challenges), stored at fuelling stations, and pumped into vehicle tanks not unlike natural gas. It can be burned in conventional combustion engines with minor modifications, but the better way is to use it through a fuel cell to produce electricity and run an electric motor.

According to Dr. David Layzell at the University of Calgary, the oil companies are looking at hydrogen with a certain amount of trepidation:

It's making the oil companies both nervous and interested. In Canada, natural gas currently sells for between $1.50 and $3 a gigajoule. Canada's crude oil attracts a price of $6 to $10 per gigajoule (equivalent to $37 to $60 per barrel) and a wholesale price of about $19 to $20 per gigajoule when the oil is upgraded to gasoline and diesel fuel. Therefore, it is clear that society places a higher value on transportation fuels like gasoline and diesel than on heating fuels like natural gas.

Now consider that transportation-grade hydrogen fuel can be made from low-cost natural gas with very low carbon emissions at a wholesale cost of about $10 to $14 per gigajoule. In other words, a zero-emission transportation fuel—hydrogen— can be made today at a lower cost than the current wholesale price of diesel or gasoline!

Moreover, using a fuel cell, a gigajoule of hydrogen will move a vehicle up to twice as far (sometimes more) than a gigajoule of gasoline or diesel using an internal combustion engine.

Many natural gas companies see low-carbon hydrogen as an exciting new market opportunity, and many oil companies see it as a threat to their existing market for transportation fuels.

Currently, about 70 million tonnes of hydrogen are produced every year. Most of it is used as "industrial feedstock" to make things like ammonia fertilizer or to upgrade and refine crude oil. Very little hydrogen is made to be used as a fuel, but that may change.

Hydrogen is mostly a by-product of the natural gas industry, and it comes in three "colours." Grey hydrogen is made from natural gas through a process called reforming, where high-temperature (700 to 1,000°C) steam causes a chemical reaction that produces carbon monoxide, carbon dioxide, and hydrogen. The carbon monoxide is then converted to more hydrogen and carbon dioxide. That CO_2 and contaminating gases are removed, leaving pure hydrogen. This is done regularly on an industrial scale and requires energy to perform, and greenhouse gas emissions are produced in the process. This hydrogen is not completely clean.

More than 90 percent of hydrogen production today is grey.[5]

Blue hydrogen uses the same process, but the carbon dioxide produced is captured and sequestered underground. However, most carbon sequestration today is used to push more oil out of a well, called enhanced oil recovery, which is not clean either. Permanent carbon storage involves extra cost with no payback, so unless it is subsidized by the government, or less expensive than paying a carbon tax, fossil fuel companies have little incentive to do it.

Finally, there is green hydrogen, which uses electricity from renewable energy sources such as wind, solar, or hydro to break water, or H_2O, into hydrogen and oxygen through electrolysis. This method is the cleanest of all. Hydrogen produced from alternatives such as wind can also be used for energy storage when the wind does not blow. However, battery storage, which is more efficient, has become the method of choice for alternatives.

Even though the production of hydrogen from fossil fuels, plus delivery and storage, involves carbon emissions, the total greenhouse gases are still half that of burning petroleum directly. Plus, the use of hydrogen in transportation can cut petroleum use by 90 percent.[6]

In the transition to net-zero-emission transportation fuels, one way for oil sands companies to stay in business may be to "gasify" the oil sands to produce hydrogen and CO_2 rather than producing crude oil that would be converted to gasoline and diesel. By capturing and geologically storing the CO_2, the resulting new energy system would have very low emissions.

We are doing the techno-economics on this now, and our results to date suggest that making low-carbon hydrogen from oil sands may generate better economic returns than traditional oil sands production, especially in a world of rising carbon taxes.

Dr. David Layzell

But there is another method to extract hydrogen from oil: do it while the oil is still in the ground.

Proton Technologies in Saskatchewan, Canada, uses the first patented system that injects oxygen into a depleted oil well that also contains water underground. The oxygen sets up a naturally occurring form of spontaneous combustion that heats the oil above 500°C. Since the oil is deep underground under pressure, it does not burst into flame; instead, the interaction between the super-heated water and the oil releases hydrogen while the carbon dioxide remains underground. A by-product of the process is the heat, which can be used to produce steam to generate electricity and run the system or can be sent out for district heating. The process is not entirely new. The oil industry has used it to help oil flow more easily out of the wells; they just didn't pay much attention to the hydrogen because the value was in the oil. Proton Technologies uses a special membrane to extract the pure hydrogen with no carbon emissions and is attempting to show that the process is commercially viable.[7]

Thousands of depleted wells dot the land, many containing 70 percent oil. That's a lot of hydrogen waiting to be released from the ground. The same technique, where the carbon is left in the ground, can also work on oil sands and coal beds. All of this is done using existing technology on oil wells that have already been drilled, and in some cases, abandoned.

Another way to get energy out of oil while it is still in the ground is to let microbes do the work for us. Certain microscopic organisms eat oil. They take care of oil spills such as the massive Deepwater Horizon disaster in the Gulf of Mexico. A lot of that oil sank to the bottom and was unrecoverable by surface ships. But over time, microbes in the sea water broke down some but not all of the oil. They also take care of oil that leaks up from the seafloor in natural oil seeps.

Underground reservoirs are no exception when it comes to microbes, so one idea is to send down ionized metal ions, which are molecules that are missing atoms. The microbes use the ions to oxidize the oil, producing electron-rich components that can be brought to the surface and run through a fuel cell to produce electricity. The idea is to basically turn the oil reservoir into a giant battery.

Early experiments with microbes have not produced enough electricity to be commercially viable, but it's an interesting idea and shows innovative thinking when it comes to ways of getting energy from oil without burning it directly.

————

Of course, the biggest obstacle to a hydrogen economy is a lack of infrastructure to move it around and store it, and of technology to use it. Following the energy crisis of the 1970s, there was a competition between hydrogen-powered fuel cell vehicles and battery-powered electric cars. The hydrogen offered greater range, lighter weight, and faster fill-up times. The race was similar to VHS versus Betamax video recorders in the 1980s. (For young readers, back in the Stone Age before home computers and streaming video, movies were stored on reels of plastic tape that came in cassettes and were available for rent in stores like Blockbuster.) In that case, VHS won in popularity, not because it was technically superior but because it offered longer playing times and was available across a number of different brands (Betamax was only available through Sony).

Today, electric vehicles have pulled ahead of hydrogen largely because of the efforts of companies like Tesla that have put tremendous research into improving battery technology and installing a wide network of charging stations. Fully electric vehicles are now produced by the millions with all major car manufacturers offering new models. Batteries are constantly being improved to provide greater capacity and faster charging times, while charging stations are popping up over wide areas; it's now possible to drive across much of North America in an electric car. Hydrogen refilling stations, on the other hand, are much rarer and more expensive to install, with fuel cell–powered cars few and far between.

Hydrogen will most likely appear first in fleet vehicles, such as delivery trucks or buses, that return to a central depot each evening, so only one filling station is needed for all of them. It can also be used in industrial processes that currently use natural gas and for marine shipping.

There is no question that oil and its consumption in the current form will continue for some time: there are simply so many cars on the road and so many support industries that keep them running. But eventually, the demand for oil will run out as gasoline- and diesel-powered vehicles go the way of the steam engine and the horse-and-buggy.

Hydrogen is going to play a role in the energy transition. We estimate hydrogen will provide up to 30 percent by 2050 if we want to meet the climate goals. That will require an eight-fold increase in the hydrogen we make today in Canada. Today, the hydrogen is made to be an industrial feedstock. In the future, we will also make hydrogen for use as a fuel, and the amount needed will be seven or eight times the current production.

We're going to have to figure out how to get there from here. We are going to see a big increase in intermittent renewables, but you can't store electricity very well. There is a role for hydrogen in the mix.

As we develop a new energy system, it is important not to think in silos, the way we've been thinking about energy systems in the past. We have an opportunity to develop an energy system that is more balanced, smarter.
Dr. David Layzell

Critics of hydrogen say it takes more energy to make than the energy you get out of it. It does take more energy to make, compress, transport, and distribute hydrogen than it does natural gas. But it takes energy to make any fuel. Consider everything involved in making gasoline. Oil must be drilled from deep underground and transported thousands of kilometres through pipelines to refineries, where hydrogen is used to "crack" the molecules and heat distills the oil down to lighter elements, and then that gasoline is shipped all over the country, by pipeline to depots as well as by truck to filling stations. That's a lot of energy being used in a process that is heavily subsidized by governments.

If the oil industry wishes to remain viable, a long-term vision might be to think of itself as an energy producer, whether of hydrogen or, through some alternative way, a form of energy from oil that leaves the carbon behind. This is an opportunity for innovation, creative thinking, and imagining a clean energy future that includes oil.

Saturn V *Apollo* moon rocket lifts off. Upper stages burn liquid hydrogen and liquid oxygen fuel.

© NASA

Chapter Eleven

Hydrogen Power

It is a super-clean fuel, but how do we make, store, and transport hydrogen?

In the same way that the ominous image of a mushroom cloud over Hiroshima gave nuclear power a bad name, the sight of a flaming ball of burning hydrogen rising above the airship Hindenburg has tarnished the reputation of hydrogen.

The picture of the world's largest airship exploding in mid-air above Lakehurst, New Jersey, in 1937 and the heart-wrenching broadcast of radio reporter Herb Morrison who could barely contain his emotions as he witnessed the terrible event have left an indelible mark in aviation history.

"And oh, it's . . . burning, oh, four or five hundred feet into the sky. It's a terrific crash, ladies and gentlemen. The smoke and the flames now and the frame is crashing to the ground, not quite to the mooring mast. Oh, the humanity and all the passengers screaming around here. I told you. It's—I can't even talk to people whose friends were on there. It—It's . . . I—I can't talk ladies and gentlemen."[1]

This tragic incident is what many think about when the topic of hydrogen power comes up.

"Won't a hydrogen-powered vehicle explode like the Hindenburg?"

According to Dr. David Layzell, professor and director, Canadian Energy Systems Analysis Research (CESAR) Initiative, University of Calgary, hydrogen is actually safer than gasoline.

> The Hindenberg was more than eighty years ago. All our fuels we use today are dangerous; they are energy dense, and even electricity is not safe if you don't handle it properly.
>
> We've learned over the last eighty-four years how to handle dangerous goods. We have rules, regulations, and standards, and many people would argue that hydrogen is safer than many of the fuels we use today.
>
> If you were to think about coming up with gasoline today, and you say, okay, I've got this gasoline stuff and it is really explosive and burns really well. We're going to put it into a gas pump and you can take the nozzle off a machine at the side of the road, put your credit card in, and spray the gas all over the place—our fueling systems today would never be approved if they were introduced today.
>
> Hydrogen is incredibly safe. The tanks are so much more robust. If it does leak, it comes out as a jet; the hydrogen diffuses incredibly quickly since it is a very small molecule. If it catches fire—absolutely, it's a torch, but it's a torch in one direction. It doesn't spread out and catch your car on fire. So, I would argue that it is a safer fuel than gasoline and diesel.

In reality, a hydrogen-powered car has less chance of going up in flames than one powered by gasoline. Even though hydrogen is a highly flammable gas, it is lighter than air, so if there is a leak, it rises and then dissipates quickly into the atmosphere, whether or not it is burning. Gasoline, on the other hand, is a liquid that falls downward when there is a leak, where it can pool under a vehicle and turn it into a barbeque. That's why fire crews are quick to respond to car accidents because there is a good chance there will be a gasoline leak and fire. Power boats driven by gasoline engines are equipped with blowers to remove heavier-than-air gas fumes that can gather down in the bilge after refuelling. Many boats have been blown out of the water by operators who start their engines without using the blowers. Hydrogen does not gather in low places like that.

Yet still, because of the Hindenburg, hydrogen is perceived as dangerously explosive.

The Hindenburg, at 248 metres long, was the largest airship ever built. It was three times longer than a 747 jumbo jet and held aloft by 140,000 cubic metres of hydrogen gas. There are many theories on what caused the explosion, ranging from sabotage to a lightning strike, or perhaps a spark from control cables operating the rudder that rubbed together. The explosion was the largest hydrogen fire in aviation history. But that giant fireball, as spectacular and deadly looking as it was, is not what killed the passengers. The hydrogen fire was over in less than 45 seconds and took place above the ship, in the sky, as the light gas rose upwards.

If you look at the historic footage of the crash, you can see people running away from the wreckage. Sixty-two of the ninety-five passengers survived, along with the captain and some crew members. Many of them simply jumped through windows as the burning craft settled to the ground. The hydrogen fire was above their heads as the gas rose, and while there were burns and nine crewmen in the forward compartment were killed by the heat of that fire, survivors were able to escape by running under it. The thirty-six deaths were caused mostly by the collapsing structure and burning diesel fuel from the engines that remained on the ground and burned for hours afterwards.[2]

It's unfortunate that a single dramatic event and powerful image have permanently stigmatized hydrogen power. The airship was initially designed to carry non-flammable helium. However, during that time, most helium supplies were in the United States and pre-war political tensions between the US and Germany prevented the Germans from accessing the safer gas, so they were forced to use hydrogen. Since the disaster, all modern airships, such as the Goodyear Blimp, use helium.

Despite its scary reputation, hydrogen is a remarkably clean fuel. When combined with oxygen during combustion, or to produce electricity in a fuel cell, the by-product is H_2O, water. That's it. No carbon dioxide to heat up the atmosphere, no carbon monoxide or other poisonous gases to pollute the air and damage lungs—just plain water.

Hydrogen was the fuel of choice for the space program during the *Apollo* moon landings and space shuttle eras. The mighty Saturn V rocket, the largest machine to ever fly, had two upper stages powered by liquid hydrogen and liquid oxygen, which were the sections that carried *Apollo* astronauts to the moon and back. The giant orange tanks on the belly of the space shuttles were filled with the same fuel. Even during the *Challenger* disaster in 1986, when one of the solid booster rockets that was not running on hydrogen sprung a leak that ignited the fuel tank in a terrible explosion, it was not the hydrogen fire that killed the seven astronauts on board. At that point in the flight, the shuttle was upside down, hanging under the tank, so when it exploded the hydrogen fire continued upwards while the cabin containing the crew broke off and arced down to the ocean. Evidence found later showed that the crew was killed by the impact with the water, not the explosion itself.

So, if hydrogen is such a wonderful fuel, why aren't we using it?

The biggest issue, besides the negative public perception, is the fact that hydrogen is a fuel, not a source of energy. It is a way of carrying energy around, the way electricity is an energy mover, not an energy source. You can't drill hydrogen out of the ground the way you can oil because, even though hydrogen is the most common element in the universe, it is quite rare on Earth in its pure form. When our planet was forming billions of years ago out of an enormous primordial cloud of gas and dust, much of the hydrogen bonded with oxygen to make the water that fills our oceans, or it was incorporated into minerals in the soil. Later, after life appeared, it became part of hydrocarbons such as oil, gas, and coal. In fact, when you burn gasoline in a car, it is the hydrogen that is breaking off from the long chain of carbon atoms that is providing the heat to run the engine. So technically, gasoline-powered cars are actually running on hydrogen, which is why sometimes you will see water dripping out of a tailpipe. That's the hydrogen, H, combining with oxygen, O_2, to produce water, H_2O. But all that leftover carbon that doesn't burn forms toxic compounds in the exhaust gas, black smoke from diesel engines, and copious amounts of carbon dioxide, which is why gas and diesel engines pollute.

The important thing to remember about hydrogen is that as a fuel, an energy carrier, it has to be manufactured. Unfortunately, that takes energy, in some cases, more energy than we get back when we use the hydrogen as fuel.

Hydrogen can be produced in two ways: by extracting it from fossil fuels, usually natural gas, or by breaking apart water into its components of hydrogen and oxygen using electricity, a process known as electrolysis.

"Reforming" hydrogen from natural gas involves high-temperature steam that separates the hydrogen from carbon dioxide and other components of the gas stream coming out of the ground. That process takes energy. Most hydrogen in use today is produced this way. On the other hand, if the hydrogen comes from a fossil fuel and it takes more energy to make hydrogen than we get back when we burn it, and then you add up all the carbon emissions involved in making the hydrogen, some argue that it is better to just burn the natural gas, which is easier and more efficient. That might make economic sense, but the carbon from hydrogen production could be captured so it does not enter the atmosphere, while burning natural gas releases carbon dioxide straight into the air. CO_2 from generating stations could be captured, but not from the tailpipes of vehicles running on natural gas.

Electrolysis is a standard high school chemistry experiment: wires from a battery are immersed in water creating bubbles of hydrogen and oxygen that rise to the surface. This can be done on a larger scale using electricity from the grid, or if the electricity comes from clean sources such as wind or solar power, then the whole process is truly pollution free.

An early idea to use hydrogen for transportation purposes was to adapt the fuel systems of conventional combustion engines in cars and trucks to run on hydrogen, similar to how they can run on propane or natural gas. That way, the automotive industry could remain basically the same. While the experimental cars did run, they found that even though hydrogen carries more energy than natural gas, it is less dense (which is why it is lighter than air) so it takes up more space. That means you either need a bigger tank or you need to compress the hydrogen quite a bit to fit it into a smaller tank. And because hydrogen burns at a lower temperature, engine power was reduced by 15 percent or more and the vehicles had to consume much more hydrogen to do the same job. Finally, since air is mostly nitrogen, which is also being consumed in the combustion process, one of the by-products of burning hydrogen directly in an engine is toxic nitrous oxide.

There is, however, an opportunity to incorporate hydrogen into large diesel engines such as those in trucks and ships. Blending it with diesel fuel can make the engines run more efficiently and cleanly.

> We're trying to get up to 60 to 70 percent hydrogen. When a big truck is under load, such as when accelerating, you might go down to 10 percent hydrogen, but when it is running at a constant 90 kilometres per hour, just running down the highway, you can get 60 to 70 percent. When the truck is idling, we can go to 85 percent.
> Dr. David Layzell

Unfortunately, you can't just blend the hydrogen into the diesel fuel like an additive. The trucks will require a separate tank for the hydrogen, adding to the complexity and cost of the vehicle. So, for the moment, hydrogen can be run mostly in larger stationary engines, such as power generators.

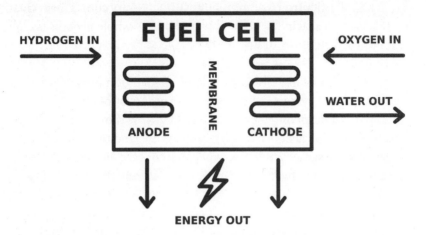

Cross-section of a fuel cell showing the flow of electrons.

© Tasha Vector Adobe Stock

A more efficient method to run vehicles on hydrogen is through fuel cells. Think of a battery with a fuel tank attached to it. Regular batteries hold all their electric charge within that black brick of a container under the hood of a car, the black slab in your device, or shiny cylinders in flashlights. If you want more power you need to add more batteries, and those batteries need to be regularly recharged. A fuel cell produces electricity like a battery, with a positive and negative terminal, but rather than have the energy stored in the chemicals within the battery, the energy is stored in the hydrogen, which can be continuously supplied. The hydrogen enters the cell and is mixed with oxygen from the air to make electricity and water.

How a Fuel Cell Works

A fuel cell is a remarkably simple device with very few moving parts. Like a battery, it has a positive and negative side, but running between them is a proton exchange membrane, or PEM. Hydrogen is pumped into the anode, or positive side, and air, which contains oxygen, is run through the cathode. A catalyst, such as platinum, strips the electrons away from the hydrogen, leaving positively charged protons. The membrane in the middle only allows the protons to pass over to the negative side. The electrons want to reunite with their protons, but they can't pass through the membrane, so they have to take a longer route around the outside through wires to reach the other side. That flow of electrons from one side of the fuel cell to the other is the electric current that is put to work.

When the electrons do reach the other side, they rejoin with the hydrogen protons to form H_2, which joins with the oxygen in the air to make H_2O or water. If the hydrogen is flowing, the electric current is too. In a vehicle, the fuel cell may operate an electric motor directly, or it can be used to charge smaller batteries that run the vehicle. A vehicle powered by a fuel cell car is two to three times more efficient than a combustion engine, but not as efficient as a fully electric car.

Fuel cells have been around since the early days of the space program where they were used to supply oxygen and water for spacecraft. Since rockets run on hydrogen fuel anyway, supply was not a problem.

Many experimental fuel cell cars and light trucks have been developed, with Toyota, Hyundai, and other manufacturers producing road-ready cars that have been introduced to Japan, California, and England. These are basically electric vehicles with the battery pack replaced by the fuel cell to provide electricity. California has pledged to make all its public transit vehicles zero emission by 2040. Numerous hydrogen fuelling stations have been set up in Los Angeles and San Francisco as well as in Vancouver, BC, to service the cars. The big advantage of hydrogen over fully electric vehicles is that the tank can be filled in a matter of minutes rather than the half hour or so needed to charge batteries.

Currently, the biggest obstacle to hydrogen vehicles is the availability of refuelling stations, which are much more expensive to set up than electric charging stations. Hydrogen is a more difficult fuel to transport and store: because it is such a small molecule, it easily leaks from conventional containers and pipelines.

If, on the other hand, the hydrogen was turned into liquid form, the situation would be very different. When gases are condensed down into liquid form, they shrink in volume by about four hundred times, so you can get a lot more of it into a smaller space.

If we could make liquid hydrogen on heavy trucks, ships, and trains, and you had a submersible pump in the liquid hydrogen, which would pump the liquid hydrogen out of its storage container so it expands into a gas just before it goes into a fuel cell. Trucks could drive 2,000 kilometres without refuelling. Or we could send a train from here (Calgary) and refuel it in Toronto.

You don't need that many refuelling stations when you look at liquid hydrogen.

Dr. David Layzell

Liquid hydrogen is a wonderful idea but there are two catches: first, liquifying hydrogen is an expensive industrial process and, second, the liquid product is -252°C. That is insanely cold, which means it must be kept in a well-insulated tank under pressure to keep it from boiling away into a gas. Also, handling a liquid that cold will require highly specialized and expensive equipment.

Watch historic footage showing a close-up of a Saturn V moon rocket launch and you will see large chunks of ice falling off the sides of the rocket during liftoff. That was moisture from the air freezing against the ultra-cold liquid hydrogen fuel tanks. The big external fuel tanks on the space shuttles were orange-coloured from a foam material that was sprayed on the outside for insulation. It was a piece of that foam breaking off from a tank during launch of space shuttle *Columbia* that punched a hole in the wing and doomed the mission to break up in the atmosphere as it returned to Earth.

For practical purposes, the hydrogen used for vehicles will be stored as a compressed gas rather than a super cold liquid. Think of exchanging the gas tank on a car with something similar to a scuba tank.

Even though the technology for the vehicles themselves has been ready for decades and even though hydrogen is an incredibly clean fuel, the difficulty and expense of storing and distributing hydrogen is the main drawback to fuel cell vehicles for consumers.

One way around the problem is the electrolyzer, a refrigerator-sized unit that you can use at home to make your own hydrogen to power your fuel cell vehicle. Australian company LAVO has developed a unit that hooks into a home plumbing system and uses electricity, preferably from solar panels or wind turbines on the roof, to make hydrogen from water. To make the system safe for home use, the hydrogen is absorbed into a porous metal hydride that acts like a sponge. This technique holds much more hydrogen, and is safer, than if it were stored as a gas. The unit can easily release the hydrogen and run it through a fuel cell to produce enough electricity to run a home for a day. This technology can be scaled up to provide energy storage for wind and solar farms when the wind doesn't blow and the sun isn't shining.[3]

> When it's a windy day, you've got so much electricity you don't know what to do with it, and the price of electricity drops because it is a free-market bidding system.
> This creates an opportunity for hydrogen generation. If the grid price is above, say, $30 a megawatt hour, you put electricity on the grid. If the price is below that, you make hydrogen, opening up an opportunity to grow our wind and solar resources. Wind and solar power is the cheapest electricity that we can make, much cheaper than electricity from natural gas.
> Dr. David Layzell

Hydrogen has seen its most success in fleet vehicles such as buses that return to the same central depot for refuelling. Canadian company Ballard Power Systems operated fuel cell buses in the city of Vancouver during the 2010 Olympic Winter Games as a demonstration project and is working with major European and Californian cities to convert their bus fleets to hydrogen. Another Canadian company, New Flyer, has developed a fuel cell hybrid electric bus where batteries run all the systems, including the drive motors, with the hydrogen fuel cell providing the electricity to keep them charged. Since the fuel cells keep the batteries topped up all the time, the long wait times for battery recharging at the end of the day are eliminated. Only the hydrogen needs to be refuelled, which can be done in a matter of minutes.

In 1998, Iceland declared that it would become the first country to move towards a hydrogen economy. The island has to import all of its fossil fuels, making them very expensive. However, since the island is an active volcano, much of the electricity is generated by geothermal energy, which taps into the hot rock that lies just below the surface. Their electricity is very cheap and free from emissions.

The plan was to use the abundant electricity to make hydrogen from water. Automotive companies were ready to provide fuel cell vehicles, and Royal Dutch/Shell would help with distribution. Go figure—an oil company helping a country get off oil. But the philosophy at Shell at the time was that they are an energy company, so it didn't matter if vehicles ran on gasoline or hydrogen, as long as the filling station they went to was Shell. Very forward thinking.

Sadly, the economic downturn of 2008 struck a hard blow to the Icelandic economy, so the hydrogen development has been put on hold.

This scenario would work for other volcanic islands such as Hawaii, which has an abundance of renewable energy as well as the world's largest volcano, Big Island. The state is experimenting with hydrogen production and fuel cell shuttle buses. But again, development has been slow because of the cost for infrastructure.

Hydrogen held great promise in the 1990s because back then batteries were large and heavy, and could not provide much range. But battery technology is evolving rapidly to improve energy density and range so it is now comparable to gasoline power. Along with a large network of fast charging stations that reduce charging times, fully electric vehicles are dominating the market for personal cars.

Personally owned vehicles, which a person uses 4 percent of the time, are parked 96 percent of the time on some of the most expensive land in Canada. If we continue to have that kind of mobility system, it should be plug-in electric.

Where you can't use plug-in electric, or where you need fuels that are more energy intense without the weight, or you need longer distances, or you can't afford the refuelling

Airbus hydrogen aircraft concept.   © Airbus Industries

time—waiting five hours to refill a vehicle—then hydrogen is the way to go.
Dr. David Layzell

One huge segment of the transportation sector that could benefit from clean technology is aviation. Emissions from airplanes worldwide amounted to just over one billion tons of carbon dioxide in 2020, which is about 2.5 percent of total human-produced carbon dioxide. But decarbonizing aviation is a big challenge.[4]

European aircraft manufacturer Airbus Industries has unveiled plans to develop hydrogen-powered zero-emission airliners to reduce the industry's impact on climate change.

Advances in engine design and improved aerodynamics have made today's jets, like cars, far more efficient than they were in the 1950s. Also like cars, there are many more of them than there used to be. To avoid increased emissions, and to cut current emissions, the industry will need to find another way to fly.

Airbus's solution is to burn hydrogen instead of kerosene-based jet fuel in the engines. When hydrogen combines with oxygen during combustion, the product is water, so while there will be no carbon emissions, the engine exhaust could produce extra-long contrails (condensation trails).

Airbus's new ZEROe concept is a plan for three aircraft configurations: a turboprop commuter, a mid-range turbofan jet, and a futuristic blended-wing airliner. All will use engines that have been designed to run on hydrogen. The company calls the planes "hydrogen-hybrid" aircraft since an electric powered boost will also be provided by hydrogen fuel cells.[5]

Liquid hydrogen can deliver four times the energy of jet fuel, at one-third the weight, which is a big advantage in aircraft. Unfortunately, while it's light, it's not very dense. So even supercooled liquid hydrogen takes up four times more space than jet fuel to deliver the same amount of energy. That means reimagining how to store fuel on the plane.

Normally, jet fuel is stored in the wings, but the new Airbus plan anticipates large tanks within the fuselage at the rear of the plane. In other words, within the same body that carries the passengers. This might make some people nervous knowing there is explosive fuel behind their seats.

But pre–Second World War zeppelin catastrophes aside, hydrogen is inherently no more risky than any other fuel. Airbus's new concepts include a venting system that runs up through the tail to disperse any fuel leakage away from the plane. The lighter-than-air hydrogen rises up and dissipates very quickly. A leak of jet fuel or gasoline, on the other hand, pools under the aircraft and burns from there. There have been many cases where passengers have survived an air crash, only to die in the fire that breaks out after the plane hits the ground.

One major difficulty is that hydrogen in liquid form is super cold, which makes it more difficult to handle. This means the fuel tanks must be well insulated, and equipment such as pumps and pipes must be designed not to freeze up when in contact with such cold liquids. This will require a huge transition at airports, with new fuelling procedures and storage systems for cryogenic fuels.

Hydrogen has been around for decades, driving the early space program and running fuel cell cars and buses. But it has remained somewhat on the sidelines because of a lack of storage and distribution networks—plus the fact that batteries have become so much more efficient and inexpensive. But batteries are still too heavy for larger aircraft, so hydrogen offers a clean alternative to fossil fuels. If the airline industry can prove that it is practical and safe, hydrogen may work its way more fully into other areas of transportation, such as shipping, to keep vehicles moving without compromising the climate.

There is one more reason the race between hydrogen and battery power for vehicles could change in the future.

Most batteries today are based on lithium, which is in limited supply around the world. The largest deposits are in Chile with the rest scattered through Australia, Argentina, China, the United States, and Zimbabwe. As demand for batteries rises, the cost of lithium will rise. Hydrogen, on the other hand, is virtually unlimited when the source is water and emissions free when it is made using green technology and run through a fuel cell. So, in the future, we may see a renewed interest in hydrogen power as the cost of batteries reaches parity.

Hydrogen is a wonderful fuel on its own, but it comes with a lot of baggage: a complicated process to make it; difficulty transporting, storing, and handling; plus a certain public fear factor. But as we look to a clean energy future, hydrogen's time will come, and hopefully by that time, the image of the Hindenburg will be forgotten.

Gases escaping from smoke stacks could be captured
and stored underground.

Chapter Twelve

Carbon Capture

Can we store carbon underground to keep it out of the atmosphere?

Admit it—you've done it at least once. We all have . . . swept a little dirt under the carpet.

It's a quick solution to a dirty problem, and you hope it doesn't come back to haunt you.

We have dirt in our atmosphere that comes from burning our dirtiest fuel: coal. But that coal is also the most abundant and one of the cheapest fossil fuels with a long legacy of providing reliable energy, so it is hard to give it up. Even though most countries are switching to clean alternatives, the demand for coal is still high, especially in China, India, and South Asia as they build their rapidly growing

213

economies and support huge populations. Together, they represent 69 percent of world coal consumption, which is likely to continue for decades.

So, what do we do with the black soot that clogs lungs and the carbon dioxide that is warming the atmosphere? One solution is to sweep it deep underground where it will remain trapped in geological formations, otherwise known as carbon capture and underground storage (CCUS). It is also referred to as clean coal technology, which the energy industry sees as a way to keep using a cheap fuel source without the harmful emissions. According to the International Energy Agency, 40 million tons of CO_2 were captured in 2020, with more expected in the future:

> After years of a declining investment pipeline, plans for more than 30 new integrated CCUS facilities have been announced since 2017. The vast majority are in the United States and Europe, but projects are also planned in Australia, China, Korea, the Middle East and New Zealand. If all these projects were to proceed, the amount of global CO_2 capture capacity would more than triple, to around 130 Mt per year.[1]

It's a nice idea, but it is a small fraction of the 36 *billion* tons of CO_2 emitted annually worldwide. This form of capture also comes with a price.

How a Scrubber Works

The technology to capture soot and some of the gases coming out of smokestacks has been around for decades. Scrubbers use static electricity to attract the tiny particles and prevent them from leaving the stack, then they use a spray of solvents that absorb carbon dioxide. Common solvents are aqueous ammonia or monoethanolamine (MEA). Once the CO_2 is captured it is separated and sent to storage while the solvents are returned to the process. However, these scrubbers are prone to corrosion and can produce waste slurries that must be disposed of. In some cases, a by-product is gypsum, which is used in products such as drywall.

Other techniques use absorbers that attract carbon dioxide, thin membranes that let all gases through except CO_2, or cool flue gases until the greenhouse gas condenses out. While effective, all these techniques require energy, so a generating station equipped with scrubbers will put out less power.

The most common technology to remove gases and soot from large facilities such as generating stations is the scrubber. Scrubbers are effective on large operations where all the waste gases come out of one big stack. But there is no scrubber technology that can be fitted to the millions of tailpipes on vehicles all over the planet, which is one reason carbon emissions are still rising. It is astounding how an invisible gas that only makes up a tiny percentage of our atmosphere can have such a powerful effect on the global climate. Carbon dioxide makes up less than 1 percent of all the gases in our air, which is mostly comprised of nitrogen, oxygen, and water vapour. Carbon dioxide is measured in parts per million, which is a minuscule amount. If you were to drop black ink into a litre of water at 400 parts per million, the current level of CO_2 in the atmosphere, you would hardly see a difference. The water would still be clear. So how is this gas, found in such small amounts, able to turn the global climate up and down like a thermostat?

Here's how the whole process works. Sunlight does not directly heat the air. Most of it passes right through and is absorbed by the ground. This is a good thing, because if air was not transparent like glass, you wouldn't be able to see through it: the sky would be completely black, and we wouldn't see anything day or night. It would also be incredibly cold on the surface because solar radiation would not be able to penetrate the atmosphere and heat the ground. It is only when light is absorbed that its energy is turned into heat, so it is the ground that gets warm and that heat, in the form of infrared radiation, is released back up into the air.

You can feel this effect on your body during the first warm days of spring when we get the first break from winter's grip. It feels warm when you are in the sun, but as soon as you step into shade or the sun goes behind a cloud, it feels cold. That's because the ground is still frozen and not radiating much heat. What you feel is the direct radiation from the sun warming your body. It is only after the ground has thawed and warmed up that we get the scorching days of summer.

Our atmosphere is basically heated from the ground up, which is one reason temperatures drop as you go higher. You are getting further away from the heat source. (Air also expands as it rises, which causes cooling.)

It turns out that the carbon dioxide molecule, which is one carbon atom and two oxygen atoms bound together, can vibrate at a frequency that resonates with infrared light, otherwise known as heat. If a beam of infrared hits a CO_2 molecule, the light energy is absorbed, causing the molecule to ring like a bell. Eventually, the molecule will stop ringing and release that infrared heat back into the atmosphere, where it can excite other molecules or be absorbed as heat into the ground. Because the heat did not escape into space, and the sun continues to add heat to the planet, there is a feedback loop where the amount of heat coming in is greater than that going out so the temperature continues to rise. That is the greenhouse effect. Other molecules, such as methane, water vapour, ozone, and nitrous oxide, do the same thing.

Of course, we don't want to eliminate all greenhouse gases from the atmosphere because they do serve a purpose; in fact, they are necessary to keep the Earth's environment hospitable for life. If we didn't have CO_2 in the atmosphere, the Earth would be a frozen ice world with an average global temperature of -18°C. But thanks to those gases we are kept at a comfy 15°C. The greenhouse effect works to keep us warm at night as well, acting as an insulating blanket when the sun no longer heats the ground. That is a very powerful effect for a gas that makes up only 0.04 percent of the atmosphere.

By contrast, on the moon, where there is effectively no atmosphere, the ground is heated to 127°C during the day and drops to -173°C at night.

Generally, there is a balance between the amount of sunlight being absorbed by the Earth and the amount of radiation leaking back out into space. Not all sunlight reaches the ground. Some is reflected back by white ice and snow, or the white tops of clouds. Not all infrared radiation is trapped by CO_2. A lot of it passes back up through

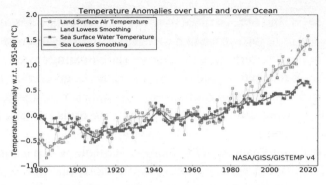

Graph showing temperature rise of land and sea for the last 140 years.

Image courtesy of NASA Earth Observatory

the atmosphere into the cold depths of outer space, the ultimate heat sink. When that balance exists, the average temperature of the climate remains reasonably constant.

Scientists are able to trace the history of the Earth's climate from actual samples of ancient air recovered from bubbles in ice cores drilled out of glaciers. These bubbles date back hundreds of thousands of years revealing how carbon dioxide levels have changed in the past and how the average temperatures have gone up and down with it. There have been cold ice ages, and then, during periods of intense volcanic activity when vast amounts of CO_2 were added to the atmosphere, the planet has been warmed well above what it is today. The most extreme was the end of the Permian Period 250 million years ago, when volcanoes raged for thousands of years, raising the average temperature 10 to 30°C above current levels while the oceans became highly acidic. Together, these effects caused the largest mass extinction in the history of Earth, where 95 percent of life was extinguished, including that in the oceans.

At the other end of the scale, when CO_2 levels were lower, the Earth passed through five ice ages, going back a million years or so. There is even geological evidence suggesting a super-ice age 650 million years ago called the Snowball Earth, where the entire planet, right down to the equator froze up. (There were other factors affecting ice ages, such as the position of the continents and ocean circulation, but CO_2 was still a factor.)

Between the ice ages there have been interglacial warm periods lasting about 10,000 years, and we have been fortunate to be living in one of these Goldilocks zones, where the climate has been not too hot and not too cold for the last 12,000 years. Based on this pattern of hot and cold cycles, we are overdue for another ice age, but that is not happening thanks to human activity. We have our hand on the thermostat and are cranking it up faster than has occurred in the past, increasing levels of CO_2 in decades by amounts that would have taken centuries or millennia in the past. And that is the concern. It is not so much the amount of CO_2 in the air, because the Earth has been warmer in the past, but the rate of change. A rate so fast that the environment struggles to adapt.

So if this pesky little gas is the source of so much trouble, all we need to do is get rid of the excess, right?

Since carbon dioxide is spread around the entire globe, there are two approaches: suck it directly out of the air or try to catch it at the source. It is like a person lighting up a cigar in an elevator full of non-smokers. Either a fan with a filter could be turned on to suck out the smoke or, the better solution, the person could be asked to extinguish the cigar.

The best solution is to stop putting carbon dioxide into the atmosphere in the first place. The best sources to go after are smokestacks in coal-fired generating stations and industries such as cement production, steel, transportation, and heating that produce a lot of CO_2. The idea is to capture the carbon dioxide before it leaves the stack and store it away permanently.

Carbon capture has been used by the fossil fuel industry for some time, but the captured CO_2 is pumped underground for what is known as enhanced oil recovery. This is where oil wells have given up their easy-to-reach supply, so the carbon dioxide is used to force more oil out of the well. Since that new oil will eventually be burned, there is no gain. However, this method can have a positive effect because some of the CO_2 remains underground. If that amount exceeds the amount of carbon that would be released by burning the oil, there can be a reduction in emissions to the atmosphere. Some of that underground CO_2 mixes with the oil to help it flow better, and if that is separated from the oil and returned underground in a closed loop, the amount of storage goes up. But there is still CO_2 going into the atmosphere.

An alternative approach is to treat the fossil fuel before it is burned and turn it into another type of fuel. When gasoline is heated in pure oxygen (autothermal reforming) or put through high-temperature steam (steam reforming), the products are carbon monoxide and hydrogen. The carbon monoxide is reduced with water to produce carbon dioxide, which is removed, and the hydrogen goes on to be burned on its own as a fuel. This process involves extra costs, although now that hydrogen is being considered an alternative fuel in vehicles and some industrial processes, it has increased value.

The most effective way to dispose of carbon is to bury it deep underground where it will remain permanently. The largest example in North America is at the Boundary Dam Power Station in southern Saskatchewan in central Canada. It is the world's first commercial carbon capture and storage project. An existing coal-fired generating station was retrofitted with a $600-million addition to capture carbon and modifications were made to the plant for a total cost of $1.2 billion. The project is designed to capture 1 million tonnes of carbon per year, which the company says is the equivalent to removing 400,000 cars off Norwegian roads. The site was chosen because the region is rich in coal deposits, so the carbon capture is a way to allow the generating station to continue operating without emissions. Since opening in 2014, the plant has not met expectations due to problems with the technology but continues to operate around 70 percent capacity.[2]

Gases from the coal furnaces are directed to a separate on-site facility that processes them in a somewhat complex system to remove the CO_2. According to Dr. Arvind Rajendran at the University of Alberta:

> In the traditional method, you take the flue gas, which is primarily nitrogen, CO_2, and oxygen, that comes out of your stack and put it through an absorber, a liquid that contains an amine molecule, NH_2 molecule, which reacts with CO_2. The liquid only takes up the CO_2. The nitrogen and oxygen go out into the atmosphere.

You take the liquid that contains the CO_2, heat it up, which releases the CO_2 in a fairly concentrated form, and then you can recycle the liquid. You use the liquid in a loop and through this you can increase the purity of the CO_2 to 95 to 99 percent.[3]

Since the process involves heat, there is a cost.

You produce steam in a power plant, so the idea is that you use some of the steam to heat up the fluid to remove the CO_2. That means you are using part of the steam that could have gone into making electricity, which means the power plant has to produce more to capture the CO_2 that it produces. That is called a parasitic energy, which is about 30 percent. So that is a loss to the power plant because you have to capture the CO_2. The energy penalty is about 30 percent.[4]

This is another case where the technology is clean, but the process is not cheap.

In addition to factoring in Boundary Dam's cost of building the facility and the energy penalty to the electricity generation, after the carbon dioxide has been concentrated and compressed, it is sold to Cenovus Energy, an oil company that injects it underground for enhanced oil recovery. Again, since that new oil will be burned, some of those 400,000 cars will be put back on the road.

Since the unit came fully online in 2019, more than 4 million tonnes of carbon have been captured. The gas that does remain in the reservoir down below in natural rock formations that held the oil in place should keep the CO_2 trapped indefinitely, although no one knows the long-term effects and whether that gas will eventually leak back out.

Another project called Shell Quest, operating in Canada's famous oil sands project, has captured more than 5 million tons of CO_2 since it began operating in 2015 and stored it permanently underground in sandstone rock formations. While that is a positive step in the right direction, that storage still only represents one-third of the CO_2 emissions from Shell's operations on that site.[5]

If a power plant is not located near a suitable geological formation, but happens to be along a coastline, there is another option to store the carbon dioxide. It can be pumped into the deep ocean. If the gas is delivered more than 3,000 metres down, the extreme pressure and low temperature of the water will keep the carbon dioxide in a liquid form, which should remain at depth. However, there are no long-term studies on how long it would stay down there, whether it would slowly dissolve into the sea water causing acidification, or what effect it would have on deepwater marine organisms.

Fossil fuel powered electrical generating stations are the usual targets for carbon capture and storage. But they are not the only industry that can be targeted.

> If I burn natural gas, it only contains 4 percent CO_2. If I burn coal, the flue gas contains 15 percent CO_2.
> If I look at a cement plant, the CO_2 is 20 or 30 percent.
> As a chemical engineer, one of the first things we teach our students is the amount of energy that you have to spend is proportional to the concentration jump you need to have. So going from 4 percent to 95 percent is much more difficult than going from 30 to 90. We have to think: what is the stream that I am recovering this from?
> *Dr. Arvind Rajendran*

The manufacture of cement produces a ton of carbon dioxide for every ton of product. About 4 billion tons of concrete are used worldwide every year.[6] Cement is made from powdered limestone that is heated and mixed with other ingredients to form a product called clinker. Forty percent of the weight of limestone is CO_2, which is released during the process. That's a lot of CO_2 going into the atmosphere. In total, the manufacture of cement represents 8 percent of global greenhouse gas emissions. That's more than aviation produces.

Several companies are now producing cement with carbon dioxide as one of the ingredients in the cement itself. CarbonCure uses industrial CO_2 and injects it into the process so it mineralizes and becomes calcium carbonate, similar to limestone. This actually makes the concrete stronger.[7]

CarbiCrete Technology replaces some of the clinker with ground steel slag, a waste product of the steel industry. The CO_2 binds with the slag granules. At this time, it can only be formed into blocks, not poured. The company claims a kilo of carbon is stored with every block.[8]

Solidia makes cement at a lower temperature, then cures it in CO_2 so it becomes incorporated into the cement, saving 70 percent of emissions.[9] Again, this process is only for precast blocks and could be used as foundation blocks or even railway ties.

If these products became widely available, there would be a demand for CO_2, which would provide incentive for power companies to invest in carbon capture. Interestingly, the construction industry is dedicated to lowering their emissions, which they acknowledge are appallingly high. Storing carbon in building materials, rather than using it to enhance oil production, is a strong alternative to keep carbon out of the atmosphere.

An interesting variation on mineral storage of CO_2 uses rocks. Experiments in Iceland injected CO_2 into volcanoes where it reacts with basalt rocks to form carbonate, minerals similar to limestone. The heat of the volcano speeds up the natural process, but it still requires large volumes of water, 25 tons for each ton of CO_2 buried. This technique will only work in volcanic regions close to a coastline. Then again, if it is close to the sea, why not pump the CO_2 into the deep ocean and save all that water?

To help drive innovation in this process, the XPRIZE Foundation, which was responsible for aiding the development of the first private spacecraft, SpaceShipOne, has now put up a $100 million prize that "aims to develop breakthrough technologies that will convert CO_2 from power plants and industrial facilities into valuable products such as building products, alternative fuels and other products we use every day."[10]

The competition will last through Earth Day 2025.

And speaking of building materials, there is also a move back to wooden structures, which naturally store carbon. We can also rewild

areas by planting more trees, enlarge conservation areas, and prac-
tise regenerative farming to plough CO_2 into soils.

One advantage to capturing carbon from power plants and industry
is that all the carbon dioxide is in point sources, right there at the
bottom of a smokestack. But what about all the carbon that has
already been put into the atmosphere since the beginning of the
Industrial Revolution? Trying to capture that takes us back to the
smoker-in-the-elevator analogy: if you can't convince the person to
put out the cigar, you can hope that a fan might remove the smoke
and clear the air.

Large-scale direct air capture device.   © Climeworks

   Believe it or not, there is a project underway to do the equivalent in the atmosphere by sucking carbon dioxide directly out of
the air. The technology, called direct air capture, uses a series of
huge fans. At a proof-of-concept pilot plant in Squamish, British
Columbia, air is brought in through a series of closed-loop chemical reactors that not only remove the carbon dioxide but also recombine it with hydrogen to produce synthetic fuel that can be burned
in vehicles. Vehicles become carbon neutral because any CO_2 they
put into the atmosphere will be captured and recycled again. Engineers estimate that industrial-scale versions could remove a million
tonnes a year, but since the world currently releases more than

30 gigatonnes of carbon a year, it would take thousands of these around the world to scrub the air.

Direct air capture, going from 4 percent in a natural gas plant to 400 parts per million, is one hundred times more dilute.

There are two things that come into the picture. The first is the energy required to do any separation. In fact, it goes approximately by the logarithmic scale of the concentration, which means for one hundred times more dilution, you spend three to four times more energy.

The second issue is that if you're going to pick needles from haystacks, you have to process the entire haystack to pick the needles. That means you have to do this on very large scales, the equipment has to be very large.

The other thing is that you don't need to put it close to a point source, like a power plant. You can put it close to a storage site. But you need energy to do that, and that has to come from somewhere—ideally, a non-fossil fuel source.

People have proposed to position these close to nuclear plants or geothermal plants because they have waste heat that doesn't produce CO_2 that could be available. A Swiss company has two big projects, and they do less than 1,000 tons per year. Compare that to a million tons from Boundary Dam.
Dr. Arvind Rajendran

One important note about this technology is that if the product is fuel to be burned in vehicles, it is carbon neutral, not carbon negative. That means the amount of carbon in the atmosphere is not reduced; it essentially remains at the same high level it is today. Only permanent carbon storage lowers CO_2 levels in the atmosphere.

These technologies are also very expensive. The $1.24-billion total retrofit of the Boundary Dam project was for just one out of four burner units, which is why very few power plants have adopted the technology. It comes down to a question of who will pay for those upfront costs: the company, the taxpayer, the customers?

———

A recent paper published by the Institute of Physics found that of all the carbon capture and storage projects around the world, they will not be enough to meet targets.

> Strengthened climate goals and new investment incentives are delivering unprecedented momentum for CCUS, with plans for more than 100 new facilities announced in 2021. CCUS technologies will play an important role in meeting net zero targets, including as one of few solutions to tackle emissions from heavy industry and to remove carbon from the atmosphere. Although recent progress is encouraging, the planned pipeline of projects would fall well short of delivering the 1.7 billion tonnes of CO_2 capture capacity deployed by 2030 in the Net Zero by 2050 scenario.[11]

Among the problems was making a sound financial case. When carbon is captured then used for enhanced oil recovery, there is a valuable product at the end. If the carbon is permanently stored underground, there is considerable cost but no sellable product. Permanent storage is just for the altruistic goal of slowing down climate change. Another obstacle, at least in the United States, was government regulations and policies that made it difficult to obtain drilling permits, which slowed the process down and increased costs. In the end, the researchers found that for carbon capture and storage to work on a large scale, it will need regulatory barriers to be brought down, government incentives such as tax breaks and loans, along with industrial buy-in using proven technology.

According to the United Nations Intergovernmental Panel on Climate Change, if we are to meet our 2°C target by 2050, we need to not only reduce carbon emissions but also go below pre-industrial levels by drawing out 8 billion tons of CO_2 from the atmosphere every year till then. Carbon capture and storage could accomplish that, but to do it, most smokestacks would have to be converted, air capture machines will need to be built around the world, and underground storage of CO_2 will have to be true storage and not a means to enhance oil production. This will involve tremendous upfront costs, which makes it difficult to enlist investors.

Of course, the simplest solution might be to simply stop burning fossil fuels. But that is not going to happen any time soon.

Unless you are doing an aggressive move to renewables, in most reasonable scenarios, it looks like you will still be producing a lot of CO_2. If we are to reach net zero, we need technologies that will bring down the CO_2, and that is a place where I see CO_2 capture fitting in, at least as a bridge for the next thirty to forty years. And going forward, unless we find ways of making chemicals from sources other than fossil fuels, we will need these options to go to net zero.
Dr. Arvind Rajendran

Carbon capture is an end-of-pipe, or end-of-smokestack, solution. While it feels good to stand before a generating station and talk about all the carbon that was prevented from going into the atmosphere and the equivalent number of cars that were removed from the roads, we have to look at the other end of the process. We have to ask where all that carbon ends up. If it is truly stored away permanently, that can be a good thing. But if it goes to pumping more fossil fuels out of the ground or being used for fuels, the benefits to the atmosphere are not as great as they might seem.

More research is needed to bring the costs of true capture and storage down to reasonable levels until clean, cheap alternative ways of generating electricity, such as wind and solar, come online and we can retire those fossil fuel generators and their carbon capture for good. The cigar in the elevator needs to be snuffed out.

City lights shine brightly along the east coast of the US in this
photo taken from the International Space Station.

Chapter Thirteen

Energy Efficiency—The Invisible Power Plant

*How much energy can be saved by making homes and
buildings more efficient? What are the upfront costs
and payback times?*

The absolute favourite room for astronauts aboard the International
Space Station is the cupola, a circular pod with seven large windows
resembling the nose cone of the Millennium Falcon from *Star Wars*.
Poking straight down from the underside of the orbiting complex, the
cupola is the only place where crew members can enjoy a 360-degree
panoramic view of the Earth, 400 kilometres below. Astronauts spend
hours photographing their home planet, identifying landmarks, mountain ranges, the red of deserts, the blue of the oceans, all streaked with

white clouds. But at night, they get a completely different view, where cities shine like galaxies in the darkness—New York, Paris, Singapore, London, wherever humans gather in large numbers. Lights shine upwards with a luminescence that decorates the Earth with glimmering ornaments.

It's a beautiful sight. It is also a sign of how much energy we are consuming, and how much we are throwing away. Unlike the campfires thousands of years ago, our modern fires are visible from space. It prompts a challenge from Amory Lovins of the Rocky Mountain Institute: "I say to integrative designers, go forth, be fruitful, and subtract."

© NASA

We are all familiar with the iconic "Blue Marble" view of the whole Earth taken by *Apollo* astronauts from the moon in the late 1960s. It was the first time that we, all of humanity, got to see ourselves from afar as a small somewhat-fragile blue planet floating in a very large black space. That photograph, along with the book *Silent Spring* by Rachel Carson, about the deadly effects of chemical pollution, kicked off the environmental movement. We all became aware that the Earth is a single living organism, a closed ecological system where changes in one place affect everything else.

Those pictures of the Blue Marble Earth were single frames from cameras held by human hands. In fact, the very first "Earthrise" picture was not even planned. "It was not part of our itinerary," explained Bill Anders, former astronaut, who took that first picture as a crew member of *Apollo 8*, the first mission to orbit the moon. I had the fortune of sharing a breakfast with Bill where he noted, "Our mission was to photograph the moon and document future landing sites, not look at the Earth."

Apollo 8 was the first time that humans left the Earth and committed themselves to the gravitational pull of another heavenly body. They did not land on the moon's surface, but they were the first to see the whole Earth all at once. Bill went on to explain:

> We were on the third orbit and the nose of the command module happened to be pointed forward as we came around from the far side of the moon, and suddenly there it was—the beautiful blue Earth rising above the lunar horizon. It was beautiful. The commander, Jim Lovell, said, "Did you get the shot?"

Indeed Bill did, making it one of the most iconic pictures ever taken.

While the view of the Earth today from the cupola of the International Space Station is spectacular, modern astronauts are not afforded that full view. They are too close to the planet to see the whole Earth in one shot. Only the twenty-four men who went to the moon had that privileged view. But as dramatic as it was, the view from the moon showed mostly the dayside of the Earth. So, in 2012, NASA began assembling a series of satellite photos of the Earth at night into what became known as the "Black Marble." These full-globe, cloud-free images clearly show the continents outlined in the light from cities. North America, Europe, and Southeast Asia all shine brightly, while Africa remains almost entirely dark. The images are updated periodically to show the growth of cities as humans continue to spread our influence across the face of the globe. They also track the spread of wildfires, gas flares, and changes in areas of human conflict.

We have become so good at lighting up the night side of the planet that astronomers are finding it more and more difficult to find dark skies where their telescopes can peer into the cosmos.

While hauntingly beautiful, and favourite targets for astronaut photography, the lights of Earth as seen from space are the most visible representation of the enormous amount of energy we use, and how much of it we waste.

Most lighting fixtures are mounted on ceilings or tall poles and point downwards onto our feet so we can see where we are going. Any light that shines upwards and leaks out into space is light we are not using, in other words, wasted energy. So those beautiful pictures taken by astronauts are evidence of our inefficient use of energy. And we waste a lot. Oftentimes we throw away more than we use. And it is not just light. We throw away heat, we throw away water, we throw away food and many products, like plastic, that are only used once.

As we look for ways to reduce our impact on the environment and lower our demand for energy, a big step towards that goal is to cut down on waste by improving the efficiency with which we use energy. Think of it as a virtual power plant, a resource we can tap into by lowering our energy consumption. We save money along the way and even provide jobs.

Building Efficiency

In a previous chapter, we talked about the inefficiency of the internal combustion engines driving most vehicles today, where only about 20 percent of the energy in the fuel actually gets to the wheels. The other 80 percent is thrown away as waste heat and greenhouse gases. But there is another less obvious source of carbon that has been standing before us all along and is responsible for more than one-third of city emissions: buildings.

According to the Intergovernmental Panel on Climate Change:

In 2010 buildings accounted for 32 percent of total global final energy use, 19 percent of energy-related greenhouse gas emissions (including electricity-related), approximately one-third of black carbon emissions.

In existing buildings 50–90 percent energy savings have been achieved throughout the world through deep retrofits.

Energy efficient appliances, lighting, information communication and media technologies can reduce the substantial increases in electricity use that are expected due to the proliferation of equipment types used and their increased ownership and use.[1]

How many movies have opened with an overhead shot of a city looking straight down on skyscrapers, usually over Manhattan? What you really see are rooftops covered with large air conditioning units, big fans, and ventilators, all spewing waste heat and carbon emissions into the air. So much so that cities develop their own heat islands, huge bubbles of hot air that can create their own microclimate. That is all energy we are throwing away, energy that could be put to better purposes.

Fortunately, the situation is improving thanks to regulations regarding insulation for homes and new buildings. According to Amory Lovins, energy expert with Rocky Mountain Institute, "saved energy is already the world's largest source of energy services, bigger than oil (i.e., 1990–2016 reductions in global energy intensity saved more energy in 2016 than the oil burned in 2016)."[2]

With structures built during the 1930s, '40s, and '50s, a time when energy costs were low, if you needed more heating or cooling, all you needed to do was add a larger furnace or air conditioning system. Meanwhile, the buildings themselves were so poorly insulated that much of the energy leaked through the walls out into the atmosphere. Capturing the heat before it escapes and containing it within the walls, plus taking advantage of the energy falling out of the sky through passive solar, is a powerful way to reduce our carbon footprint. And the beauty of energy efficiency is that it provides jobs in the renovation industry and usually pays for itself in less time than expected.

RETROFITTING

A surprising success story of a deep retrofit is the iconic Empire State Building in New York City. The 102-storey art deco tower, built in 1931, stood as the tallest building in the world for nearly four decades and is still one of the most recognizable buildings today. Thanks to improvements completed in 2010, energy use in the building has been reduced by 40 percent, resulting in cost savings of $4 million annually. Ten years after the retrofit, the value of the savings had already exceeded the cost of renovations and is expected to pay for itself several times over.

The amazing part of this story is that the retrofit did not change the look of the building and was remarkably simple to achieve. The first step was to install more than six thousand reflective barriers behind each radiator used to heat the rooms. Before the barriers, about half of the heat the radiators produced was going into the poorly insulated walls then outside to heat up New York City.

Next was the replacement of the tower's 6,500 windows with super-insulated, gas-filled double-pane units with extra sealing around the edges. The windows were coated with an insulating film to block ultraviolet rays from the sun for the occupants' safety and to limit excess sunlight during hot sunny days. This more than tripled the insulating value of each window. To keep room temperatures constant, automated blinds controlled by an intelligent system opened and closed according to the amount of sunlight. Low-energy LED lighting was installed throughout with automated timing systems to reduce lighting overnight and on weekends.

The combination of improved windows and efficient indoor lighting reduced the heating and cooling demands on the building by one-third. It meant smaller, more efficient, less costly heating and air conditioning units could be installed.

Even the sixty-eight elevators running up and down the tower contributed to energy savings. Regenerative braking systems capture the momentum of the cabins on the way down and as they slow to a stop, producing electricity that is fed into the building.

A huge energy saver came from the tenants who occupy the building. Each is provided with a digital dashboard that displays their energy consumption of lighting and electricity. It turns out that people are more likely to reduce their energy consumption when they can see how much they actually use.

The total retrofit of the Empire State Building is estimated to cut carbon emissions by 105,000 metric tonnes over fifteen years. There are plans to reduce energy consumption a further 30 percent by 2030, making it a beacon of energy efficiency that will light the way towards New York City's Climate Mobilization Act. This is a plan to impose emissions caps on all buildings in the city over 25,000 square feet (2,323 square metres) by 2030. It is the largest carbon reduction plan of any city in the world. If every commercial building in New York followed the same blueprint, carbon emissions would be reduced by 4 million tonnes, roughly the equivalent of one coal-fired generating station.

GREEN BY DESIGN

Retrofitting older buildings is, of course, only part of the story. More and more, engineers and architects are striving to design new buildings that meet the highest standards of efficiency. The international standard is LEED (Leadership in Environmental Energy and Design), the most widely used green building rating system in the world. It has inadvertently set up a healthy competition among designers to come up with the most efficient buildings in a city, country, or even the world. Efficiency not only saves on energy costs. There is pride and prestige involved for both the builders and the people who occupy the buildings who earn bragging rights as part of such an impressive clean and green building.

Close to the top of the list for North America is the Manitoba Hydro Place in Winnipeg. It was the first building to achieve Platinum LEED and utilizes a 115-metre-tall solar chimney that rises up the side of the building to manage air circulation. The chimney is painted black, so it absorbs solar radiation that heats the air inside the chimney. The rising air acts as a huge natural fan that provides natural circulation through the building. Movable shutters on the outer walls control the amount of sunlight streaming in. The energy required to run the building is 70 percent less than commercial buildings of comparable size.[3]

One of the top designs on the planet is the International Renewable Energy Agency headquarters in Abu Dhabi, which has won numerous awards for sunlight management and solar voltaic power supply.

The potential savings from improved energy efficiency are huge, and it can all be done using existing technology. Amory Lovins, of Rocky Mountain Institute, has been promoting this idea for decades.

It is the largest, "least expensive, most benign, most quickly deployable, least visible, least understood, and most neglected way to provide energy services" and offers expanding returns not just through mass-produced widgets but also by substituting brains and information for hardware.[4]

PASSIVE SOLAR BY DESIGN

Improving energy efficiency goes beyond trying to capture the energy trying to escape; it also involves taking advantage of the free energy coming in from nature, especially from the sun. All you have to do is change the shape of a building. It's called passive solar, and it can drastically reduce the need for fossil fuels using existing technology.

When you look at new housing projects, most of the homes look the same, basically a box with a front door, smallish windows all around, and a garage. True, new houses are better insulated than they were in the past, and solar panels or solar tiles can be installed along with a battery power wall, but there is still much more that could be done if they were designed with a different shape.

In the 1980s I was invited into a custom-built home that was designed to beat the energy crisis back then through passive solar heating and cooling, which means it captures solar energy without the use of complicated fan-driven ventilation systems. In fact, the house has no furnace at all and was built using conventional materials in Canada, where winters are very cold.

From the outside, the most noticeable feature of the rectangular house is large windows on the entire south side rising two storeys high to the flat roof that slopes towards the rear. In the centre, behind the windows is a masonry pillar about 2 metres wide that is both a support structure for the roof and an air circulator. The outside of the stone structure is painted black and covered with the same type of double-pane glass as the main windows. The simple structure is known as a Trombe wall and is very effective at controlling the movement of air through a building.

As sunlight is absorbed by the black wall, it is turned into heat that is trapped behind the glass—the greenhouse effect. That heat is stored by the stone masonry and slowly released into the house.

Black tubes attached to the wall carry water that becomes the hot water supply.

The clever effect of the Trombe wall is how it moves air. Hot air naturally wants to rise. At the top of the wall, vents leading into the house allow the hot air to enter at ceiling level, while other vents at the bottom draw cooler air from the floor into the wall to replace it, setting up a natural air circulation pattern within the house. Fresh air can also be drawn in from the outside of the wall at the bottom and heated before it enters the house.

Inside, sunlight streams through the south-facing floor-to-ceiling windows that have blinds to control the amount of sunlight coming through. It is warm in this open concept space even when it is quite cool outside. The floor is made of concrete, which was poured extra thick to absorb heat, so it feels warm underfoot and retains that heat after the sun goes down. If you have crossed a sidewalk barefoot on a hot sunny day you have experienced the heat storage capacity of concrete. Pipes running through the floor circulate hot water from the solar-powered water heater to help the floor of the house provide gentle heat at night and on cloudy days.

The most ingenious innovation was how the house could use heat from the sun to provide cooling. This is done by taking advantage of the north-facing shady side of the house, which had no windows and was partially buried by a berm, or grass-covered mound of dirt. Burying the back of the house underground helps with insulation in winter and cooling in summer. The sloping roof comes almost down to grass level with vents that open to the inside of the house just under the eaves. On hot days, these vents on the backside are opened. At the same time, vents at the top of the Trombe wall at the front of the house are opened to the outside air while those at the bottom of the wall on the inside of the house also remain open. Now, a different circulation pattern is formed. The rising air in the Trombe wall creates a suction that draws air in from the vents on the shady backside of the house. The air entering from the back is cooled when it passes over the grass on the berm, so when it enters the house, it naturally flows down along the floor to the front of the house, enters the Trombe wall where it is heated by the sun, rises to the top, and is expelled outside. Presto, natural air conditioning powered by the sun, without the use of electric fans, pumps, or refrigeration coils. Amazing.

The incredible part of this project is the fact that this super-efficient home was made from regular off-the-shelf construction materials. No exotic technology was involved. The energy savings were all due to the design of the building. Everything was integrated to work with everything else to maximize efficiency.

Again, according to Amory Lovins:

Efficiency reduces capital costs, not by adding fancy widgets, but by using fewer and simpler widgets—taking out unneeded stuff.

Efficiency doesn't deplete a limited resource such as oil. It depletes stupidity, a very abundant resource.[5]

There is a lesson here for urban planners, as well as homeowners, to rethink the concept of the family home, to think "outside the box," away from the little boxes that all look the same to efficient shapes that capture the free energy of the sun and use it wisely. Look for energy demonstrator homes in your area, step inside, and feel the difference.

Passive solar heating is not a new concept. It goes back to wartime when fuel was scarce and designers came up with ways to capture energy from the sun with things such as windmills and rooftop solar water heaters. Today, through retrofits of older buildings and reconsidering the design of new ones, the savings can bring us a big step closer to a low-carbon economy. And the beauty of it is that any upfront investments are usually paid back in a short time just in fuel cost savings.

Our cities and towns are bathed in solar energy, which is largely underutilized by all the roofs and windows in buildings and homes. Most emission reduction gains can be made from large buildings, such as commercial and residential towers above five storeys, shopping malls, big-box stores, sports complexes, warehouses, and universities. A simple retrofit such as insulation, which we normally associate with keeping heat in, is also effective in keeping heat out and reducing the cost of air conditioning.

On-site renewable energy such as solar panels on the roof, along with small wind turbines, especially in buildings that cover large areas, like shopping malls and box stores, can make the buildings self-generators. And if you consider the tremendous area of all those rooftops in a city being covered in solar panels, they could replace land that would be taken up by large-scale solar farms.

RESOURCEFUL BY DESIGN

Finally, there is an opportunity to save energy by taking advantage of untapped resources that may be in your own backyard. The city of Toronto, on the shores of Lake Ontario, took advantage of cold water at the bottom of Lake Ontario and uses it to cool downtown buildings. Enwave is the world's largest Deep Lake Water Cooling system, which uses 40 kilometres (25 miles) of pipe to serve 180 buildings, including hospitals, commercial offices, condominiums, data centres, and hotels. The system replaces expensive cooling systems that each building would need on its own, saving 55 megawatts of energy from the grid every year. That's enough to power eight hospitals.

Capturing energy from the sun or the Earth and plugging holes in walls will not solve all our energy and climate problems, but it is a big step that is relatively easy to take. It is also an opportunity for creative architects to design super-efficient buildings that will not only manage energy but also incorporate greenhouses, atriums, and other indoor green spaces where plants can take advantage of all that extra sunlight streaming in and make the indoor environments more liveable.

Transporting Efficiency

On a broader scale, there is much that can be done to improve the efficiency of cities as a whole, especially when it comes to transportation.

Endless lines of cars clogging multi-lane highways is the classic image of Los Angeles, the home of the car culture. After the Second World War, owning a car became a symbol of success, which, combined with a new continent-wide interstate superhighway network, became the ultimate freedom. As prosperity blossomed, more and more cars packed the roads to the point where that very freedom evolved into congestion and entrapment during traffic jams.

Following the post-war baby boom, cities, especially those in North America, grew based on a California model. The idea was for people to live away from the downtown core in suburbs, sprawling clusters of homes. To go shopping, take the kids to school, go out for dinner or entertainment—all involved driving. This fit perfectly with California, where the car was king.

Roads dominate all environments, and with construction of the vast US interstate highway system, one of the largest single engineering projects in the history of civilization, along with the Trans Canada Highway, people were encouraged to take driving holidays. This was a tremendous boon to the automotive industry, oil production, and tourism. No one thought about the gases coming out of all those millions of exhaust pipes because the concept of air pollution was not really recognized. North America is so big, with people spread out over such large areas, polluting gases simply blew away on the wind. Out of sight, out of mind. Or as the saying goes, "The solution to pollution is dilution."

But now the car culture and design of transportation systems is being reconsidered to reduce the impacts that exhaust gases are having on climate change.

The first step is to replace the gasoline-powered engine with electric drive, powered either by battery or hydrogen fuel cells. According to Dr. Curran Crawford at the Institute for Integrated Energy Systems, University of Victoria in British Columbia, it's about integration of all the different forms of transportation.

> I've heard the saying that a good transport policy is a good public planning policy. You need city densification and proper planning so you can support it.
>
> There is growing recognition of active transport, if you look at the advent of electric bikes that is enabling some amount of shift, although there is some evidence that it is not actually shifting people out of cars and onto buses because you can ride when you want.

But tied into that is infrastructure like bike lanes. It really comes down to making it convenient and attracting people to either choose mass transport or active transport. All the electric vehicles now are essentially skateboards with the platform, the batteries, and the motors. You can put whatever you want on top of that. This can include ride hailing variants and shuttle buses to augment central bus lines.

With more electric platforms, you can make more tailored solutions, making it convenient so you don't have to wait for the large single-bus routes.

So, let's take a look at the future of different types of urban transit, starting from outside the cities and working towards the core.

SELF-DRIVING CARS

Personal automobiles have dominated our transportation, offering independence, freedom, and symbols of status and personal expression. It will be difficult to tear those freedoms away from people, so cars will be with us for a while. However, the way they are controlled will change dramatically.

Artificial intelligence, along with 360-degree cameras and other terrain sensors, are bringing cars closer to self-driving. Already, cars are equipped with lane keeping abilities, adaptive cruise control that maintains a safe distance from other vehicles, self-navigation, and parallel parking functions. All these automations still require a human driver with hands on the wheel to monitor the driving. Fully autonomous vehicles requiring no driver, and even no steering wheel or controls, are not available because the technology still has to prove itself safe for everyone, vehicle passengers and pedestrians. According to the Traffic Injury Research Foundation, more than 90 percent of automobile accidents are caused by driver error, not mechanical problems with the vehicle.[6] From an engineering point of view, if the human driver is removed, cars will be far safer, drive at the most efficient speeds, and take the shortest route to a destination.

There are five levels of self-driving cars with capabilities increasing along with level number. Some of these features are already available.

Level 1 is basically cruise control for speed and distance, but the driver is still in full control of the vehicle. Level 2 is advanced driver assistance systems (ADAS), which can control steering and acceleration, but the driver must have hands on the wheel and can still take control any time. Level 3, or conditional driving, involves environmental awareness, so the vehicle can perform manoeuvres such as changing lanes and passing a slower vehicle, but again, the driver must be present. Level 4 is high driving automation where the vehicle is actually driving and navigating itself through traffic, but the driver can still take over if the system makes a mistake. Level 5 is fully autonomous with no steering wheel or controls. Currently, the only vehicles of this type are relatively slow pods that act as shuttles or taxis in restricted areas. Eventually, it could come to the point where personally owned cars are not necessary or, according to Dr. Anthony Perl, professor of Urban Studies and Political Science at Simon Fraser University, not desirable.

> If everyone has an autonomous electric car, it's not going to work too well for energy and sustainability purposes.
>
> If, on the other hand, you have the equivalent of ride hailing, where you just press a button on your smartphone and a vehicle shows up, maybe with one or two other people in it, and it takes you where you want to go, that works because the amount of space and energy use starts to make sense. We have to learn how to share our spaces and our mobility. There is a place for autonomous electric vehicles but not to duplicate the two-car garage with all of them privately owned.

The technology for self-driving cars is advancing faster than regulations and public acceptance. Current federal regulations do not allow Level 5 fully self-driving cars on the highways until the technology can be proven safe. And from a marketing point of view, an important consideration is whether people will trust driverless cars over their own driving skills. Before fully autonomous, hands-free cars are allowed on the road, there are some other specific barriers that must be overcome.

First is the issue of liability. Are computers capable of making the right decisions in an accident situation? If a driverless car becomes involved in a fatal collision, who is responsible—the owner or the manufacturer of the machine?

Then there is the human factor as driverless cars are integrated into traffic surrounded by cars with drivers. Let's be realistic: many people push the law a little by driving slightly above the speed limit or racing away from stop lights to show off all the horsepower under the hood. Automated cars, on the other hand, will drive exactly according to the rules of the road, which may seem annoyingly slow to those who are in a little more of a hurry. It might lead to a new kind of road rage and even violence against those cars, where angry drivers in older, scratched-up vehicles will give the new machines a little nudge on the highway just to test how good they really are.

During the transition to self-driving cars, when they do share the roads with humans, there will be dedicated lanes where only self-driving cars are allowed, such as the current HOV lanes on expressways. In these lanes, the computer-guided systems will enable the cars to travel extremely close to each other, less than a metre apart, and maintain that distance at very high speed. This would dramatically reduce commuter time and improve traffic flow. For commuters, it means worry-free driving, time to read the headlines, catch up on e-mail, or connect with international partners rather than wondering what the road-rage idiot in the lane beside you is going to do next.

If self-driving cars do take over the roadways completely, those who still want to enjoy the thrill of driving themselves will have special parks where they can drive their vintage cars around tracks on weekends. But they won't be allowed to take them on the road.

But wait a minute. If we are going to line up vehicles closely together and run them at high speed across the land, that technology already exists. They're called trains.

Personally owned cars do not really belong in cities. They take up parking space, are a danger to pedestrians, and, frankly, become a burden to the driver once the destination is reached. Cars belong in outer areas where public transit is not practical or as transportation to travel to remote locations. If you want to go downtown, take a self-driving taxi to the train station.

Japanese Shinkansen high-speed rail.

HIGH-SPEED RAIL

Rail is the most efficient form of land transport when it comes to the amount of energy required per ton of cargo. If you look at a convoy of trucks rolling down the highway, each unit has its own engine and the load is supported by eighteen wheels mounted on soft rubber tires made from fossil fuels and likely to wear out within a year. The soft tires present a lot of rolling resistance with the asphalt highway, also made from fossil fuels, that must be constantly repaired and resurfaced. A train can pull more than a hundred cars with two or three engines that use steel wheels running on steel rails with very little rolling resistance. Trains are up to six times more fuel efficient per ton of load than trucks and emit a similar amount less carbon.[7]

Cars emit roughly 192 grams of carbon per passenger kilometre, while diesel passenger trains emit 41 grams and an electrified train only 4 grams.[8] The world speed records for passenger vehicles on land are held by electric trains, with the French TGV the fastest on wheels, reaching 574.8 kilometres per hour (357.16 miles per hour) during a special experimental run in 2007.[9] In April 2021, France banned some internal air routes in favour of trains to reduce the country's carbon emissions.[10] The Japanese Shinkansen, or bullet trains, run at 320 kilometres per hour in regular service;[11] other high-speed rail systems operate at similar speeds in China and other countries around the world.[12]

> It is a very mature technology. It's been around for over fifty years. There have been two waves in Asia and one in Europe. Japan introduced the technology for the Tokyo Olympics in 1964. People thought that maybe it can only work in a place that is very densely populated and very linear, like Japan.
>
> But then about twenty years later, France, Germany, Italy, and other European countries, even the United Kingdom with the Channel Tunnel, established it on that continent. More network possibilities evolved with hub and spoke in Paris and Madrid and medium-sized networks in Germany and Sweden.
>
> The third wave arrived about ten to twelve years ago in China, and the rest of the world over the past decade. It showed that a very large nation with a big population can operate at scale nationally. About 80 percent of China's population has access to high-speed rail, which improved their domestic mobility. Unfortunately, the country doesn't use enough green energy yet, but they are moving in that direction.
> *Dr. Anthony Perl*

With high-speed rail so well established around the world, is it possible to go even faster at ground level? It is—if you remove the wheels altogether.

The absolute speed record holder for trains is the new Chuo Shinkansen Maglev, which has reached 603 kilometres per hour (376.8 miles per hour) on a test track in Northern Japan.[13] The train is held a few centimetres above the track by magnetic levitation, which is the repulsive force you feel when you try to push two like poles of magnets together. The maglev train uses two different sets of magnets: one to levitate the train, the other to keep it away from the walls of the U-shaped travelway. A set of coils embedded in the track propels the train forward. With no contact with the ground, rolling resistance is much lower so extremely high speed can be achieved.

How Maglev Works

Think back to the first time you encountered magnets. You were probably amazed at how they would stick to certain things magically without glue. You were probably even more amazed when you tried to put two magnets together and they pushed each other apart. You might have also played a trick where you placed a metallic object like a paperclip on a table, held a magnet under the table, and made the object magically move around by moving the magnet underneath.

These two principles of magnetic repulsion and movement are what drive a maglev train.

The simplest way to magnetically levitate a vehicle would be to have magnets on the underside with, say, their north poles facing down, and another set of magnets embedded in the track with their north poles facing up. The vehicle would be suspended in the air by the repulsion of the like poles facing each other. But permanent magnets are very heavy and can't be controlled in any way. Their magnetic fields are always the same.

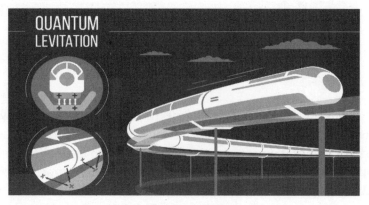

Magnetic levitation suspends the train above the track and moves it forward so there is no contact, reducing rolling resistance.

Another way to achieve the same effect is to use electromagnets. Recalling high school science class, that's where an electric current run through a coil of wire turns the coil into a magnet. The stronger the current, the stronger the magnet, and if the electric current is reversed, the polarity of the magnetic field reverses as well.

The one other effect in play is induction. When a magnet moves over a conductor, such as a piece of aluminum, tiny electric currents are generated in the aluminum with corresponding magnetic fields that will repel the magnet above.

Applying the above to the maglev: The train has specially positioned superconducting electromagnets, so the track simply needs loops of aluminum to provide repulsion to those electromagnets.

Moving the train forward requires another set of electromagnets, a series of coils laid out in a line. By sequentially alternating the currents in the coils, the changing magnetic fields flow like a wave down the line of coils, pulling the train along with it. The magnets on the train pull it towards a coil, then just as it passes over, the coil reverses field and pushes the train so it continues forward. The speed of the alternating fields determines the speed of the train. These coils can be either embedded in the track or on the train itself.

Put the two systems together, and you have "maglev," a vehicle that levitates itself above the track without touching it, glides along with no moving parts so it is extremely quiet, and has no emissions.

The drawback to maglev trains is the high cost of the track, which is two to three times higher than regular high-speed rail and sixty times higher than conventional rail. The track is also exclusive to maglev so conventional high-speed trains cannot run on it. So why is Japan going to the extra expense?

One thing people don't know about—one reason Japan is pushing ahead with its maglev—is that, after more than fifty years, they are going to have to shut down the bullet trains between Tokyo and Osaka.

Every night, they shut down the trains and do maintenance. It is one of the world's best-maintained systems. But after more than fifty years and a few earthquakes, it's not the tracks or the signals or the wires that need maintenance—they're in perfect condition. I'm sure they keep them up. It's the underbed, the subsurface, that has never really been shut down long enough to be repaved. Once the maglev is running in parallel next to it, they're going to have to shut down the bullet train for several years and rebuild the bed for the train's right of way. That was part of the tipping point for going ahead with the maglev. It wasn't because of the increased capacity; it was because they will have to shut down the bullet train for long-term repair and renovations, and they can't do that until they have something else in place.
Dr. Anthony Perl

An alternative approach, used by the German-built maglev train running from the Shanghai Pudong International Airport to downtown, involves an elevated track with the sides of the train wrapping around the outer edges of the track to the underside. The magnets use attraction to pull upwards from below. The result is the same kind of levitation, where the train is not actually touching the track so there is no rolling friction as there is with wheels. This train holds the world speed record for a regular service passenger-carrying train: 430 kilometres per hour (265 miles per hour). But it only holds that speed for a few moments on the airport route because the entire 30-kilometre trip is covered in just 8 minutes. This short line is the only maglev in China, which is rapidly expanding its high-speed train network. Why not more maglevs?

The Chinese are very good at adapting other people's technology. They bought a maglev that runs from the airport to Shanghai in the early 2000s. They didn't buy it and deploy it to save about 15 minutes to go from the city to the airport; a high-speed train could have done it for a fraction of the price. They wanted to test out the technology and see whether or not they would use it for their national system, and they decided no.

I have to think it was partly for the energy penalty. They bought French, German, and Japanese high-speed trains for different parts of their system, then reverse engineered them and created their own composite trains, which they claim are better than all the others. And they are the most successful high-speed rail operators.

The point is, if China kicked the tires, so to speak, on each type of high-speed rail plus maglev and opted for a composite of the high-speed rail technology, that shows that they didn't see—at least in this century—a long-term sustainable future for maglev.

Dr. Anthony Perl

But wait. There is one more idea to go even faster on land.

If billionaires like Elon Musk and Richard Branson have their way, they may almost double the speed of high-speed rail using a new technology called Hyperloop.

One of the biggest obstacles facing any high-speed vehicle is air resistance. At human walking speeds, air seems completely ethereal, a transparent gas we easily stride through without any effort at all. Then a hurricane shows up and suddenly we see the force of high-speed air tearing buildings apart and ripping up trees by their roots.

High-speed vehicles are creating their own hurricane-force winds as they shove their way through the air, which puts up more resistance the faster we try to go. Air resistance is non-linear, which means the faster you go, the more force is needed to overcome it. If it takes 10 horsepower to go 80 kilometres per hour, it will take 80 horsepower to go 160 kilometres per hour. In other words, if you double your speed, you need eight times more power. Eventually, you come to a point where most of your energy is spent pushing the air out of the way until you run into diminishing returns, where it takes a huge amount of energy to go just a little bit faster. That's why airliners fly just below the speed of sound. The so-called sound barrier, around 1,235 kilometres per hour (767 miles per hour) at sea level, is where shock waves form in front of the vehicle like the bow wave around

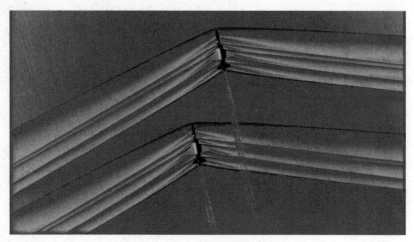

Shock waves around aircraft models in a supersonic wind tunnel.   © NASA

the front of a boat. Pushing through that shock wave to supersonic speed demands a lot more fuel, which is why supersonic flight has been mainly the domain of the military.

The only supersonic airliner, Concorde, capable of travelling at twice the speed of sound, had a successful twenty-seven-year career, but it was extremely thirsty on gas. Unfortunately, it was designed and built in the 1960s when fuel was cheap and began service in the early '70s just as the oil crisis drove the price of fuel through the roof. While Concorde could reduce travel time across the Atlantic from New York to London from eight hours down to just over three, the price of a ticket was also roughly double that of a regular subsonic flight. Concorde proved once again that time is indeed money.

One possible way around the problem is to remove the air. That's the hyperloop concept.

Imagine wrapping a maglev train and track in a tube that has had most of the air removed so resistance is effectively eliminated. And instead of a complete train, use individual, bullet-shaped pods carrying twenty-two people each. With the problem of air resistance removed, hyperloop pods are predicted to be capable of speeds exceeding 1,000 kilometres per hour (625 miles per hour). That's airline speed.

The idea of a vacuum train is not new; in fact, it goes back to 1799 when George Medhurst developed an atmospheric engine, where a

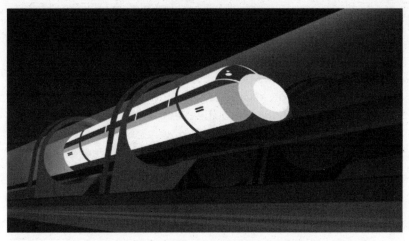

Hyperloop is a concept to carry passengers in ultra-high-speed pods running through vacuum tubes.

Image courtesy of Library of Parliament

torpedo-shaped device inside a vacuum tube embedded in railway tracks pulled a train along, replacing the steam engine. The first subway in New York was a vacuum tube for a few kilometres, and the idea has been used over the decades in futuristic science fiction. But it was Elon Musk of SpaceX fame (who else?) who revived the idea in 2013 and coined the name Hyperloop. He kept the idea open source, inviting innovations from university teams and companies around the world to develop the technology. The early concept was something he described as a cross between Concorde, a railgun, and an air hockey table. It involved a vehicle powered by a large fan similar to those used in large jet engines. The fan would suck air in ahead of the vehicle and direct it down to flat footpads that would allow the air to escape underneath providing an air cushion, like with a puck on an air hockey table, so there would be very little friction with the track. A linear motor would propel the vehicle down the tube at high speed like a pea through a straw.

The air-suspended hyperloop never materialized, but Musk held annual competitions at SpaceX where university groups from around the world could test out different design concepts. The first full-scale test run carrying humans came in 2019 when Virgin Hyperloop, sponsored by another billionaire, Sir Richard Branson, was run in Nevada: two passengers in a small pod rode through a half-kilometre-long evacuated test tube. They accelerated up to 172 kilometres per hour in about six seconds. Earlier tests without humans achieved 387 kilometres per hour. That is less than halfway to the ultimate goal of 1,000 kilometres per hour.

Will hyperloop scale up to the point where passenger pods in vacuum tubes replace airlines, or will conventional magnetic levitation without the vacuum tube prove to be the more efficient technology?

> I've spoken to people from hyperloop, and I've never gotten a straight answer when I asked how much energy is used to maintain that vacuum.
>
> Vacuums are not natural. They say, oh, we have this great sealing technology—but nothing is 100 percent and I think that either they don't really know at scale or they don't want to say. There is a cost to running a vacuum 24/7 to run these trains through. I'm sure it's less than running a maglev at 500 miles per hour at sea level outdoors, but it's not clear to me how these hyper-speed surface transportation technologies, which are technically possible, are sustainable from an energy point of view yet.
> *Dr. Anthony Perl*

If hyperloop technology achieves the jet aircraft speeds that are predicted, you will be able to travel from Los Angeles to San Francisco in thirty minutes, from Toronto to Montreal in less than an hour, with zero emissions. Engineers are predicting a capacity of thirty thousand people per hour with all pods running every two minutes on a reliable schedule that is not interrupted by weather delays.

But scaling up hyperloop to cover hundreds of kilometres and making it robust enough to handle cold climates, such as Canadian winters, is an enormous engineering challenge. The tubes will require vacuum pumps along the entire length of the course to maintain the ultra-low pressures inside; airtight seals between joints able to withstand temperature fluctuations between day and night, weathering, and seasonal extremes in some areas; and flexibility to withstand ground movements from frost and earthquakes. On top of that, the tracks inside must be built to extremely close tolerances to provide a smooth ride, which means laser straight on the flats with perfectly engineered long radius curves that have no variations. At 1,000 kilometres per hour, the pods will be covering two and a half football fields every second. At that insane speed, even a slight misalignment in the track will throw the pods and the passengers in their seats around violently.

Bloodhound SSC is designed to set a new land speed record of 1000 mph (1600 kph).

Image courtesy of Jules1982 Wikimedia Commons

Only two vehicles have travelled faster than 1,000 kilometres per hour at ground level, both specialty built British land speed cars that were run on salt flats in South Africa. Thrust SSC, powered by two jet engines out of a fighter aircraft, was the first to break the sound barrier on land in 1997, achieving 1,228 kilometres per hour (763 miles per hour). Its successor, Bloodhound, powered both by jet and a rocket, reached 1,010 kilometres per hour (628 miles per hour) on jet power alone during a test run in 2019. The team plans to push that car to 1,600 kilometres per hour (1,000 miles per hour) on a future run that will include both the jet and the rocket motor.[14]

These two highly specialized, custom-built, one-of-a-kind vehicles, and indeed any cars that attempt to break world speed records, take years of development and engineering to run at hyper-speeds, and they only maintain those speeds for a matter of seconds. Yet hyperloop promoters are planning to run the public at these speeds across great distances through vacuum tubes in pods running every two minutes.

To keep the G-forces (1 G is the gravity of the Earth, or the downward force you feel sitting in a chair) acting on the passengers to a minimum, the pods will ride up the walls of the tube on curves to keep the forces towards the floor, like an airliner banking into a turn. Otherwise, the passengers will feel like they are riding a roller coaster—fun for some, but not everyone. And because they are riding in a tube, with no visual reference to the horizon, excessive motion can trigger motion sickness.

The smallest bump or misalignment in the track would throw the passengers around violently in their seats. Canadian drivers know well the sudden jolt from hitting frost heaves that appear on our highways every winter. Imagine hitting one at 1,000 kilometres an hour!

Proposed routes for hyperloop are along flat areas such as the Nevada desert towards Las Vegas or along valley floors in California. Mountainous terrain would add considerable cost because the system would need the support of numerous bridges and tunnels to keep the tracks as straight and level as possible. On the other hand, the Japanese Shinkansen has had a perfect safety record for decades running high-speed trains through a mountainous country that is also prone to earthquakes. So, these engineering challenges can be overcome.

Canadian-French company TransPod is proposing a line between Edmonton and Calgary across 300 kilometres of very flat prairie ground, potentially reducing travel time by road from three hours to less than half an hour. The company estimates the cost to be between $6 billion and $10 billion and claims it could be economically feasible with a capacity of six thousand people per hour, with some of the pods carrying cargo.[15]

Hyperloop test tracks are under construction in the United States, Europe, and Saudi Arabia, with the first full-scale version planned for Abu Dhabi International Airport, although it will run at much slower speeds. If it works, hyperloop holds the potential to revolutionize high-speed travel, possibly replacing regional airlines with all-weather transport and low carbon emissions. Obviously, the technology will start off at low speeds and shorter distances, with higher speeds reserved for longer routes. One wonders just how far, and how fast, this concept will go, and whether it is truly affordable. High-speed rail, which is already mature technology, might be the better option.

> High-speed rail can be woven into the inner-city fabric to take people very close, either walking or a short transit ride, to where they are actually going. Airports, maglev, and hyperloop don't have that opportunity.
> There is the proven energy efficiency in high-speed rail. For its speed, it is the most energy-efficient mode that's out there, when it's fully used.
> *Dr. Anthony Perl*

STEPPING BACKWARD AND FORWARD

The final form of future urban transit is also the oldest, quietest, most efficient, and completely carbon free: feet.

Since the Industrial Revolution, cities, especially those in North America, have become more and more dependent on the car. When smoky factories were polluting the air, people chose to move away from their work, seeking cleaner country air. This resulted in large areas of land dedicated to a single purpose: sprawling urban neighbourhoods where people drive to big-box-store shopping malls, drive to theatres, drive their kids to school. By design, urban areas require driving. Living without a car is very difficult because services are so far away from homes. When you look at these commercial properties, the parking lots take up far more land than the buildings they serve. People even drive to fitness centres so they can run on treadmills!

Urban planners are now working to transform cities into walkable communities. This is really a return to the original village concept. Neighbourhoods are compact and diverse so it is possible to walk to school, be entertained, recreate, and shop—all close to home without the need for personal cars.

According to urban designer Jeff Speck, a walkable city needs four components: a reason to walk, the walk has to feel safe and be

safe, the walk has to be comfortable, and the walk has to be interesting. This can be done by getting rid of wide streets that are totally
dedicated to cars, making them narrower, with bike lanes, tree-lined
boulevards, and local shops and restaurants. Making them a place
where you can live, work, shop, recreate, and get educated.[16]

> You've probably heard the term fifteen-minute city. During the
> pandemic it is a concept that is becoming more real. I haven't
> been more than a thirty-minute walk from my house in the last
> year. I can get to everything, from a medical appointment to
> groceries to the dentist, all within that space. That is the recipe
> for a sustainable urban space. We need a lot more of that.
>
> Urban planning is essential. The idea that communities are
> complete is the exception not the rule. We need to re-allocate
> space accordingly, which means more active transportation:
> walking and cycling, more public transportation, and less park-
> ing for everyone to have a car because we won't need one all
> the time.
>
> *Dr. Anthony Perl*

Simple changes can make a big difference to the urban experience.
Merchants like it because it puts people in the streets. And those
skinny streets are safer because people drive slower than they would
on wide ones. Trees lining streets provide natural cooling in summer.

These changes are being incorporated now into existing cities
and designs for future urban areas. And it is spearheaded by the
cities themselves. The C40 Cities Climate Leadership Group is an
international organization of ninety-four global cities, including
Johannesburg, Tokyo, Cape Town, London, Beijing, Rio de Janeiro,
Paris, and Los Angeles—home to more than 750 million people and
25 percent of the world's economy. This group is showing how local
action is having a major impact on reducing greenhouse gas emissions, how together, these cities can have a global effect. One example is New York City, which has repeatedly doubled the number of
bicyclists simply by painting more bike lanes on major routes.

In a strange ironic twist, the COVID-19 pandemic, which temporarily shut cities down, helped them become more pedestrian

oriented. To help keep people safe while dining out, restaurants were given permits to open sidewalk seating. As pandemic-related restrictions ease, the tradition will continue, furthering the move towards utilizing streets on a people scale rather than giving cars priority.

The shape of cities and the way we get around in them is changing, and that will go a long way towards reducing our overall impact on a changing climate.

Improving energy efficiency is a win-win situation, where money is saved on energy costs, jobs are created with retrofits and new construction, and our impact on the environment is reduced. It may not make our Black Marble planet completely black, but it could help turn down the brightness of those lights shining in the dark and seen from space.

The ultimate zero-emission car?

Image courtesy of Fritz Ahlefeldt

Chapter Fourteen

A Great Idea But . . .

Solar updraft chimneys, air-powered cars, getting energy from dark—ideas that never made it.

Here's an idea. Mount a windmill on the front of a car so when the car moves, the windmill spins, which generates electricity that powers electric motors driving the wheels that keep the car moving, which spins the windmill more—an endless cycle that would let the car run without paying for or using any fuel.

Sounds great. Sadly, it won't work. Hundreds of ideas for perpetual motion machines have all failed because of energy losses to friction and heat within the systems. Less energy comes out than went in, and the machine remains motionless or eventually winds down to a stop. Leonardo da Vinci had it right in 1494: "Oh ye seekers after perpetual motion, how many vain chimeras have you pursued? Go and take your place with the alchemists."

Though failed ventures may abound, innovative—even outlandish—thinking is needed to develop entirely new ways to convert energy for useful purposes without harming the environment. They are certainly not perpetual motion machines, but the creative concepts presented in this chapter did try to capture energy from nature. They may have looked good on paper, and perhaps worked as laboratory experiments, but then they fall into what is known in the industry as "The Valley of Death." It begins with taking what may be solid scientific concepts and scaling them up to commercial size to try to turn a profit, and then discovering the true costs of going from laboratory models to full-sized structures: unforeseen problems that only show up when things get big, developmental delays, maintenance that pushes the cost of operating the system so high that it is no longer profitable or able to compete with cheaper existing technology—and then the descent begins. Sadly, many good ideas have fallen into this valley as investors who were enthusiastic at the beginning become nervous and pull out before the project is completed. In other cases, the ideas were too far ahead of their time and may have to wait for a new generation to pick up the idea in the future.

The Power of Hot Air

We all know that hot air rises. Sit before an open fire and be entertained by rising cinders that float almost magically up into the sky. In past times, this mysterious lifting force was called "levity" and was captured by the Montgolfier brothers Joseph-Michel and Jacques-Étienne in 1783 to achieve the first human flight. The brothers were paper makers by trade and experimented with large bags open at the bottom beneath which piles of straw were set afire. Now, a fire under a paper bag does not usually end well, especially for the bag, so they first tried wetting the bag to make it fireproof. That was about as successful a flying machine as, well, a wet paper bag.

Montgolfier hot air balloon 1783.

Image courtesy of San Diego Air and Space Museum

They improved the design by adding silk to the paper and made the world's first hot air balloons measuring 10 metres (33 feet) in diameter, which rose above the fires and hovered hundreds of metres in the sky for several minutes before gently settling down. With proof-of-concept in hand, they were ready for a public demonstration.

Before an astonished crowd of 130,00 people, including the King of France, Louis XVI, and Marie Antoinette, a basket hung from the balloon carried a sheep, a duck, and a rooster. Like the first flights into space almost two centuries later, animals were used because no one knew how humans would survive in the new realm above our heads, out of contact with terra firma.

The ornately decorated balloon standing six storeys tall rose and flew for about five minutes, landing safely a few kilometres away.

The king was impressed with the flight, although he was apparently bothered by the smell of the fire. At that time, it was believed that smoke was the lifting force, so old shoes were added to the fire to make it as smoky, and unfortunately smelly, as possible.

The big day when humans first left the Earth came on November 21, 1783, in Paris. Two passengers, Jean-François Pilâtre de Rozier and François Laurent, marquis d'Arlandes, stood on a circular platform at the bottom of the balloon and a large fire was lit below. (Imagine the nerve it took to watch a fire being lit under a giant paper bag. The interior was coated with alum for fireproofing, but still . . .) To the astonishment of the crowd, the passengers were carried aloft and sailed over Paris for about twenty-five minutes, landing softly 9 kilometres (5.5 miles) away.[1] It was a dramatic demonstration of the lifting power of heated air and took place more than a century before the Wright brothers and their airplane.

Two centuries later, a modern variation on the hot-air principle was erected in Spain, not to fly but to generate electricity.

For as long as fires have been contained in buildings, smoke has been directed up chimneys. As far back as Leonardo da Vinci, devices to capture the updraft and turn it into mechanical motion involved windmill-shaped fans that would spin from the rising hot air. Da Vinci's device was called a smoke jack and involved gears and pulleys that would turn a spit holding a pig roasting over the fire below.

Other devices were toys that spun wheels above candles, and in the 1970s patents were filed on very large solar chimneys to generate electricity. One of the more innovative designs was built in 1982, about 150 kilometres south of Madrid, Spain, near the town of Manzanares. Instead of using the heat from a fire to produce the updraft, this project used the power of the sun.

An enormous greenhouse with a flat roof covering 46 hectares (110 acres) sprouted a chimney 195 metres (640 feet) tall in the centre. It was the first prototype solar updraft tower designed to generate electricity using the power of rising hot air. The simple concept uses sunlight to heat the ground under the glass roof. The ground warms the air, which is funnelled up through the chimney. Small wind turbines mounted around the base of the chimney are spun by the passing air, while cool air from outside the structure flows along the ground through the open walls around the edge to keep the convective cell going.

The tower produced 50 kilowatts of electricity but was really just an experiment to measure the actual flow of air, temperatures during the day, etc. and never actually sold electricity. It finally collapsed in 1989 when the wires supporting the chimney corroded. Even though the tower was the height of a sixty-storey building, calculations showed that an even taller version would be necessary to be economically viable.[2] A 1,000-metre-tall full-scale version with a greenhouse 7 kilometres in diameter was planned for the Australian Outback but has yet to be built. If it is, the world's tallest building will be a chimney!

Updraft towers have an efficiency of about 2 percent, which is less than solar voltaic. More practical uses would be for them to function as actual greenhouses, or desalinization plants in desert areas where the electricity from the chimney runs the operation. So far, the large land area required and high upfront costs have kept investors away from full-scale solar updraft towers.

Air-Powered Cars

One surefire way to reduce 14 percent of global carbon emissions is to eliminate tailpipes on vehicles. Battery electric and hydrogen fuel cell vehicles are leading the way to zero emissions, but there was another approach to clean cars that roamed around the roads of India for a while: a car that ran entirely on air.

In 2007, Tata Motors unveiled its air-powered car, which used compressed air instead of gasoline to run a piston engine. Normally, gasoline and air are drawn into a cylinder and ignited by a spark. The exploding mixture expands rapidly pushing a piston down and driving a shaft that turns the wheels. A compressed-air engine does the same thing, except there is no fuel or spark. The air entering the cylinder is already under pressure and wants to expand, so the piston is still pushed down, the wheels turn, and there are zero emissions. Instead of a gas tank, the car carries an oversized scuba tank.

Airpod 2.0 vehicle runs on compressed air.

© MDI by Air Futures Group

Prototype vehicles were tested as taxis and delivery vehicles with some success, but they were found to be somewhat underpowered with an efficiency of only 25 to 35 percent, which is far less than an electric vehicle.

The problem with compressed air is that it has to be either under extremely high pressure or in very large tanks, which add weight to the vehicle. However, that has not stopped MDI, the company that invented the idea, from developing the concept further with newer light-weight vehicles, called air pods, that are ready to appear on the market under the Tata brand.

The air cars are an innovative way to remove emissions from the road; however, the energy source that drives the pumps to compress the air must be taken into consideration. If that source is wind or solar, the system is completely clean. If it is fossil fuel, then you still have cars running on oil, natural gas, or coal. But even in that case, there is some advantage because the pollution is concentrated at the generating stations, where it is easier to control with systems such as carbon capture. That is better than having carbon emissions spewing out of millions of tailpipes spread all over the country.

Air cars are a great example of innovative thinking.

Flying Autos

If travelling through a vacuum tube is not for you, perhaps you would like to take to the skies in your own quadcopter. Flying cars have been a staple of science fiction and predictions of the future going back decades. Efforts to give cars wings or make airplanes morph into cars have had only limited success, usually because the vehicles end up being compromises and not particularly good at either task.

Now, with battery technology evolving quickly and drones becoming more sophisticated, several companies are developing large drone-like vehicles capable of carrying one or two people. Of course, this conjures up the image of taking off directly from your home and flying over all that traffic congestion directly to work, the theatre, or the beach. But without some kind of lane markers or highways in the sky, it could become quite deadly to have people flying themselves in all directions at all altitudes whenever they wished. It also wouldn't be very pleasant for people on the ground with hundreds of large drones buzzing like bees overhead.

Large drones, whether carrying people or cargo, would need to be completely autonomous, like self-driving cars, so the vehicles would follow prescribed paths through the air and avoid collisions. It would eventually become an interstate system in the skies with rules and regulations for speed, direction, on and off ramps, all completely hands-free. How long will it take people to trust their lives with such a system?

Space-Based Solar Power

Here is a literally far-out idea. Imagine a satellite the size of a small town with an array of mirrors that gathers sunlight in space, concentrates it, converts it to a microwave or laser beam, and beams it down to a receiver on the ground that converts the beam into electricity. Theoretically, one of these solar-power satellites could supply the energy needs of a major metropolitan area such as New York, London, or Tokyo.

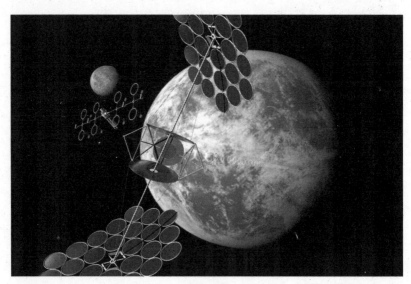

Space solar power satellite converts sunlight into microwaves and beams them to the ground.

Image courtesy of NASA

The idea has been around since the beginning of the Space Age and takes advantage of the fact that sunlight in space is four times more intense than it is here on the ground since it doesn't pass through our dusty, cloudy, turbulent atmosphere. A satellite placed in the proper orbit would receive sunlight 24/7 almost every day of the year. In the weightless environment of space, structures can be as large as we care to make them, and when it comes to solar power, size matters.

So, if these monstrosities are such a great idea, why haven't we seen them in orbit?

Part of the answer is that they *would* be monstrosities. Building something that covers many square kilometres in space is an expensive proposition requiring dozens, if not hundreds, of rocket flights to assemble. The International Space Station, which is a bit larger than a football field, took fifteen years and $100 billion to build.

The other factor is the powerful beam coming down from space. Obviously, the area above the ground receiving station would have to be a no-fly zone, and some countries might consider the device a space weapon, which could potentially start another arms race in space.

On the plus side, solar-power satellites would provide reliable, clean energy. It really comes down to who wants to pay the extremely high upfront cost of construction and how that cost, over the long term, compares to using conventional solar panels on the ground.

Space-Based Sunshade

While we're on the topic of large structures in space, how about a giant sunshade that cuts off some of the sunlight reaching the Earth to cool the climate and prevent global warming?

Since energy from the sun is being trapped by greenhouse gases and our efforts to eliminate those gases seem to be taking longer than we'd hoped, a truly out-there idea is to block the amount of sunlight shining on the Earth: use the equivalent of a giant umbrella for the planet. This device would not need to stop all sunlight—after all, we do need it for things like light, heat, and plant photosynthesis. But calculations show that reducing the amount of incoming

A sunshade in space would have to cover an area of 2.6 million square kilometres.

Image courtesy of Solar AquaGrid LLC (www.solaraquagrid.com)

radiation from the sun by just 2 percent, which doesn't sound like a lot, would bring temperatures back to pre-industrial levels. But the Earth is a very large object, so to block even that small amount of sunlight would require an enormous structure covering 2.6 million square kilometres. That's about one-third of the area of the continental United States, the area covered by the Mediterranean Sea, or twice the size of the province of Ontario.[3]

That's a rather large umbrella. It makes a solar-power satellite look like a postage stamp by comparison.

Unlike a satellite, the space umbrella cannot be in an orbit around the Earth since it would only be in front of the sun for short periods on every orbit. It must be positioned where it will always provide the shade to the sunny side of the planet, somewhere farther out and permanently between the Earth and the sun. The best spot is known as L1, which is one of five Lagrange points in space where the gravity of the sun, moon, and Earth all balance out. L1 lies directly between the Earth and the sun about 1.5 million kilometres from Earth. An object placed in a Lagrange point will remain there even though it is only a point in space.[4] (The James Webb Space Telescope was sent to L2, which is in the opposite direction, 1.5 million kilometres from Earth beyond the moon.)

Besides the logistical difficulty of building a structure the size of the Mediterranean in space, we may not actually want it to be one big piece. Something that large would be an easy target for small asteroids and comets that wander through space and could punch sizable holes in it. Over time it could be torn to shreds. Presumably

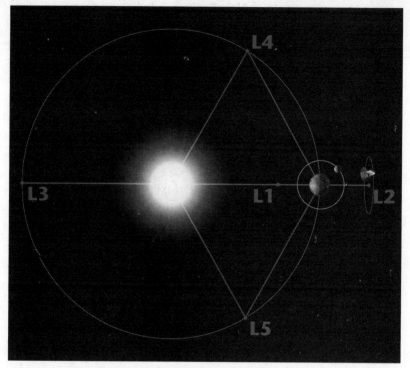

Lagrange points around sun, Earth, and moon. The James Webb Space Telescope is at L2, and the Earth sunshade would be positioned at L1.

© NASA

the shade would be made of very lightweight material, and that presents another problem. It could act as a huge solar sail that would be pushed out of position by the energy and momentum of sunlight, a concept that has been proposed for future sailing spacecraft. That would require some kind of manoeuvring system to keep the sunshade in place.

The biggest obstacle to a project of this scale is the number of rocket launches required to get all that material all those kilometres into space. Based on the carrying capacity of current rockets, it would take more than a million launches. At a cost of around $100 million per launch, we enter the realm of ridiculous.

An alternative idea is to use a swarm of smaller shades or even just a big cloud of dirt, perhaps created by towing an asteroid into position and blowing it up with a nuclear weapon. Again, the complexity of going out to an asteroid, mounting a rocket engine on it to bring it in close to the Earth, then exploding it without causing further problems is a huge challenge on its own. And even if we were able to produce such a cloud in space, if there were any unexpected consequences from the experiment that involved the entire Earth (we know that *never* happens), it would be almost impossible to clean it up.

A cheaper variation on the sunshade idea is called geoengineering, where sulphur particles are sprayed into the upper atmosphere, using aircraft or balloons, to cool the climate the way volcanic eruptions have in the past. In 1991, when Mount Pinatubo erupted in the Philippines, it spread particles around the globe and lowered the Earth's temperature by 0.6 C (1°F) for fifteen months.[5] Doing this artificially would take enormous effort involving thousands of aircraft, and would have to be maintained year after year.

The outrageous scale and cost of extreme ideas like these underlines how much simpler it is to address the climate problem at its root cause. Even the scientists working on geoengineering research admit it is a last-resort technology. It's not the sun that needs to be dimmed down, it's the excess amount of carbon dioxide we're putting into the air that needs to be lowered. If we realistically deal with the root of the problem, we won't need sunshades at all.

Underwater Balloons

Energy storage is a vital element of our new clean energy future and many innovative approaches have been tried that go beyond large battery packs. One of them is to use air pressure. Blow up a balloon and you will feel the effort it takes to inflate it to full size. As long as you hold the neck closed, the pressure inside the balloon is storing the energy you put into it. Let the neck go and the energy released is sufficient to propel the balloon around the room at high speed accompanied by farting sounds.

Now try blowing up a balloon underwater. You would be lucky to inflate it at all because water exerts more pressure on the outside of the balloon than the air you're trying to put in. You have to work harder to inflate it. That's why scuba divers must breathe compressed air underwater: it is forced into the lungs; otherwise they wouldn't be able to inhale at all since the external water pressure on their lungs would be too great.

Water pressure is another form of gravitational energy, where the weight of the water can be put to work. A pilot project in Toronto used a series of large underwater balloons that were inflated under very high pressure during off-peak times, then, when energy was needed, the pressure of the water would squeeze the air out of the balloons through turbines to make electricity. Clever idea, but working with large flexible technology underwater was a challenge. Since then the company has moved to pressurizing underground air chambers instead.

Kite Flying

It is well known that the higher you go, the stronger and more steadily the wind blows, which is one reason wind turbines are mounted on tall towers. And the amount of power a wind turbine can generate depends on the swept area, or how much sky is swept by the turbine blades. Another way to reach even greater heights and cover more sky is to fly a giant kite and take energy from the force it exerts on the string.

SkySails energy-producing kite.   Photo courtesy of the SkySails Group

SkySails Power uses an oversized parachute, similar to the rectangular chutes used by skydivers and kite surfers, to harness wind energy. Mounted on a portable device that can be located anywhere, the sail is launched to an altitude of up to 800 metres where an automatic control system causes it to gracefully fly a figure-eight pattern back and forth across the sky on the end of a tether. This arc can sweep out an area many times larger than the blades of a wind turbine using simpler technology at lower cost.

At the base, the kite tether is wound around a drum that rotates as the kite is let out. At 200 metres, a generator cuts in so that the rotations are used to generate electricity. The kite continues to pull on the tether until it reaches 800 metres, then it is positioned almost straight overhead where the tension on the cable is greatly reduced, at which point the generator becomes a motor that reels the kite back down to 200 metres where the cycle can start again. The energy used to pull the kite down is less than was generated on the way up so there is a net gain of about 200 kilowatts.[6]

A variation on the kite idea by EnerKite of Germany replaced the fabric sail with a solid wing or glider that operates in the same manner, but the flying wing is flown back down from its maximum altitude to save more energy.[7] California company Makani takes it one step further with a large aircraft on the end of a tether that flies like a kite but carries eight propellers on its wings that generate electricity as it flies. In a complicated automatic control system, the propellers are first driven by motors so that the plane hovers straight up like a quadcopter and moves away from its base as the tether is let out. Once at altitude, it transitions to sideways flight where the wind takes over so that the plane makes giant loops in the sky like a kite, with the propellers now generating 600 kilowatts of electricity. It is the first time an aircraft has flown a pattern in the sky while tethered to the ground. Unfortunately, after thirteen years of development and testing, the system proved to be too complicated and expensive compared to the lower cost of conventional wind turbines, so the aircraft/kite system never made it to commercial production.[8]

The simpler pure-kite systems like SkySails' have been proposed for remote locations where it would be difficult to build conventional turbines, for offshore applications, or for island nations. SkySails' kite system has been used to assist ships across oceans for years, proving its durability, and because it is portable, it could readily be deployed to disaster areas to provide power quickly to assist recovery. But whether a flock of giant kites waving back and forth can scale up to megawatt size and whether the public would accept them wafting over beaches are challenges that must be met.

Power from the Dark

When it comes to gathering energy from the sky, we tend to think of the shower of solar energy falling from the sun during the day. But scientists in California have developed a device that can power a light bulb using energy that flows up from the ground into space at night. This technology could fill the gap left by solar panels that only work during the day.

This new, extremely low-cost approach works like a solar panel in reverse, literally capturing energy from darkness.

Researchers at the University of California, Los Angeles and Stanford University took advantage of a principle called radiative cooling, which happens at night as the Earth radiates into space heat that it had absorbed during the day. This process can cause the ground to become colder than the air, which is how frost can form on grass overnight when the air temperature is still above freezing. That temperature difference between an object exposed to the night sky and the air is used to produce electricity.

The technology to demonstrate this concept was remarkably simple, costing about US$30. The device consisted of a Styrofoam-like box about the size of a shoebox with a metal disc painted black on the outside facing upwards and an aluminum block on the inside. When exposed to the night sky, the black metal disk would radiate its heat into space and become slightly cooler than the aluminum bar inside the box. The key component, though, was a commercial thermoelectric generator that coupled the disc and the block.

Thermoelectric generators, or thermocouples, are remarkably simple solid-state devices that generate electricity when heat passes from one side to the other. They are used in a variety of applications, from small temperature sensors to devices that produce power from waste heat flowing through smokestacks or that generate electricity from body heat, such as the body heat–powered flashlight invented by Canadian teenager Ann Makosinski in 2013. It makes electricity using the heat of the hand and ambient air so there is no need for batteries.[9]

Making electricity directly from heat has also been used on deep space probes where sunlight is too dim to provide sufficient power. Using a small piece of radioactive material to generate heat, the thermoelectric material converts it into electrical power. These devices have powered robots such as the Curiosity and Perseverance rovers on Mars, the Cassini mission to Saturn, and the twin Voyager spacecrafts that are now beyond our solar system and still sending back data after four decades in space.

It doesn't matter what the heat source is and it doesn't take much of a temperature difference for a thermocouple to produce a current of electricity. The foam box contraption to make electricity at night was placed on a roof in Stanford, California, on a clear December night for over six hours, using a temperature difference of only a few degrees between the air and the black piece of metal. It produced a peak of 0.8 milliwatts of power, which is not a lot—just a tiny fraction of the electricity a solar panel of equivalent size can produce in the day. But it was enough to illuminate an LED bulb and prove the concept that light can be produced from darkness.

The scientists say that with improvements and scaling it up, a larger device could provide enough clean electricity for lighting or charging electronics at night in a completely passive way. Thermoelectrics will not meet all our nighttime energy needs, but they could provide low-cost clean energy sources in developing countries and remote areas.

The inventions in this chapter may not have achieved the commercial success their designers hoped for, but they are examples of creative thinking and innovation. And that is exactly what is needed as we look for new ways to generate clean energy. Even the most outlandish ideas are worth a try, and if they fail to succeed, that failure can guide the way to more successful concepts.

Albert Einstein once said, "If at first the idea is not absurd, then there is no hope for it."

Image of the whole Earth taken from geostationary orbit by National   © NOAA
Oceanic and Atmospheric Administration GOES-18 Satellite.

Chapter Fifteen

Putting It All Together

What will our clean energy future look like?

The astounding aspect of green energy technology is the fact that it all exists today, most of it has been around for decades. No new inventions are required to go green. But implementing alternative technologies on a large scale and truly decarbonizing our atmosphere to prevent further climate change will take more than scientific expertise and technical skill. It will require political will, economic investment, and public acceptance. In other words, the science and technology are only half of the equation.

Moving away from fossil fuels does not mean giving up our comfortable modern lifestyle and moving back to the caves. It is a matter of subtly changing our attitudes towards energy: where it comes from and how we use it.

Our civilization has learned to adapt to environmental threats many times in the past. In 1952, the London smog brought about the first clean air act and incentives to move away from open burning of coal to cleaner technologies. The oil embargo of 1973 taught us about fuel efficiency. Brown smog over California prompted that state government to impose the strictest emission standards for vehicles, forcing the automotive industry to install catalytic converters on exhaust systems. In all cases, life in the cities went on, but the skies above were clearer.

Government incentives and environmental laws are powerful tools to promote change. Higher taxes on carbon and support for alternatives, rewriting of building codes to include more electrification, and installation of ground piping that can be used by heat pumps before the foundations are poured are driving a fundamental change in how cities evolve. But regulations can go further. Requiring commercial and residential construction to incorporate passive solar into their design to transform the little square-box houses currently sprawling across the land into buildings that are energy savers and even energy producers. Restricting personal car use in cities and, eventually, only allowing zero-emission vehicles on the streets will reduce our carbon footprint. Making these requirements law will allow cities to evolve into more pedestrian friendly places, with local services easily available, less reliance on personally owned vehicles, and greater use of regular, fast, safe public transit.

As a kid, I watched *The Jetsons*, a fanciful cartoon series set in the year 2062, where everything was in the sky. Flying cars zipped between buildings held up on high towers, while gadgets of every type made life more convenient. Some, such as video calling, actually came true. But this vision, along with earlier ones from Walt Disney and his "World of Tomorrow," have not panned out in such dramatic idyllic ways. Cities today still look like cities of the past, where people live, work, and play at ground level. But there has been a subtle evolution in technology. While cars still look like cars, now the gas tank is being replaced with a battery pack or fuel cell and the engine under the hood is an electric motor. Telephones used to be on the wall, now they are in your pocket. Our lifestyle change has not been nearly as dramatic as predicted, even though we have adopted remarkable new technology.

That is how change comes about: a slow evolution of technology that improves on what was there before rather than abandoning the old entirely and starting over with a new vision. And people love to embrace new technology.

Take, for example, how listening to music has evolved. Before 1878, the only way to enjoy music was to hear it live. But then Thomas Edison invented the phonograph that could record music on a spinning cylinder using a needle. The first version used tin foil as the surface to record, but it wore out quickly. Later he developed wax cylinders which were replaced with longer-lasting plastic versions that resembled black toilet-paper rolls and contained about two minutes of music. From that point on, professional music could be enjoyed at home. But these cylindrical shapes took up a lot of space when stored, so they evolved into flat discs, with one song on each side, that could be stacked more easily. Those grew into larger vinyl albums holding about an hour of music, then reel-to-reel tape came along so you could record your own music off the radio or try your hand at performing and recording yourself. The first tape machines were the size of small suitcases, but eventually shrank down to cassette players that fit in your hand or into the dashboard of a car. The next development in sound was the CD, which used laser light instead of needles to read optical discs, producing ultra-clean sound with no hiss or pop. Then came the MP3 player, which was just a computer chip with no moving parts, which evolved into the iPod—and today, you can store all the music you want on your phone and hear it through noise-cancelling earbuds. All that technological evolution was just to listen to music.

We have not seen that kind of evolution with energy technology because fossil fuels have been convenient, abundant, and versatile. We have built up an enormous infrastructure to keep those fuels flowing and an even larger system of technologies that burn them. Now is the time to evolve that technology into cleaner versions.

Our energy future will be a montage of different sources. The electricity running through your home may come from a local wind farm, solar panels laid across a field or on your roof, your garbage, the ground under your floor, a local small nuclear plant or new fusion reactor, tides flowing past your coastline, waves crashing on a local beach, or, more likely, a combination of many.

The beauty of clean energy is that it not only is better for the climate but also actually makes money. New jobs are emerging through start-up companies that are developing better batteries, more efficient solar cells, and larger, more powerful wind turbines as well as through the construction and maintenance requirements to get the equipment up and running reliably. Then there are the savings. As the cost of fossil fuels continues to rise, the cost of alternatives is going down. Solar electricity in many countries is now cheaper than coal so there are benefits in the long term.

There is a danger that as the cost of alternative energy comes down and more of it enters the market, the value of that energy goes down as well. It will take creative management to store excess energy when the value is low and sell it at a profit when demand rises.

The change to a new energy portfolio will come with resistance from those invested in the old ways of doing things, but rather than focus on jobs lost, there are new opportunities to be gained as the technology becomes integrated into our way of life. New products will come along, such as cheaper, more energy-dense batteries; new industries will appear, such as battery recycling to recover precious metals, the way platinum is retrieved from catalytic converters on old cars.

In the same way that fossil fuels are concentrated in certain areas around the world, such as Texas, Alberta, Venezuela, and the Middle East, so too are the alternatives. Mountain tops and flat areas have stronger prevailing winds, while deserts receive more sun. Only certain coastlines have the right shape of bays or configuration of off-shore islands to produce strong tidal currents. Sub-tropical areas are best for producing fast-growing trees. It will be up to individual areas to decide which form of energy production is best suited to their needs and resources. A lot of that energy will be distributed through smart grids, but a lot will be produced locally, freeing countries from their dependence on imported fuel.

We are in a new industrial revolution.

Will there be costs involved? Certainly. All energy capture involves costs, both economic and environmental. But as we discovered during the COVID-19 pandemic, money can be found when the need is urgent. Hundreds of billions of dollars suddenly became available internationally to address an unprecedented crisis. Funds were provided to support science that identified the virus, then an international effort launched research and development that saw laboratories around the world share information to come up with the most effective vaccines. Drug manufacturers produced millions of doses that were distributed around the globe. There was also a massive education campaign to change public behaviour about social distancing, hand washing, and mask wearing and to encourage people to get vaccinated, while support was given to business that suffered because of lockdowns. And all of that happened in less than two years.

Reaction to the pandemic was fast because we could see the effects. Seniors were dying in care facilities; loved ones, people we knew were succumbing by the thousands. Epidemiologists showed graphs of infections and deaths and told us we had to "flatten the curve." The effects of the pandemic were literally in our face. Thanks to this monumental international effort, we did better than flatten the curve: we turned it downwards.

Science has been showing us the rising curves of climate change for decades. Now it is in our face with rising temperatures, warming oceans, increased acidification, more intense storms, frequent and intense wildfires, longer droughts, shortages of fresh water. We have been told we need to flatten those curves before the climate reaches a tipping point where a feedback loop will accelerate the process and turn the Earth into a hothouse.

One example is the loss of ice in the Arctic. White ice is part of the planet's cooling system, reflecting sunlight back into space during the summer months. As ice disappears, dark sea water absorbs that sunlight and becomes warmer. This causes more melting of ice, exposing more dark areas, melting more ice, and so on in an accelerating spiral until all the ice is gone. That is why the North is warming twice as fast as the rest of the planet. Disappearing ice is only one of many feedback loops happening around the world.

And it's not just humans who are being affected by these changes. In what has come to be known as the sixth mass extinction of wildlife, animal numbers are declining due to human activity, changing habitats, dwindling food supplies, and the changing chemistry of the oceans.

While the climate scientists have been waving the red flag, a sense of urgency has been missing because the effects of a changing atmosphere are not as obvious as body bags being rolled out of senior care facilities. At the time of writing, two years into the pandemic, the worldwide death count from COVID-19 was 5.7 million people.[1] According to a recent study out of Harvard and three British universities, nine million people die every year from air pollution related to fossil fuel burning.[2] The World Health Organization estimates roughly 150,000 deaths per year from the effects of climate change, which include malnutrition and disease due to droughts.[3]

In other words, more than twice as many people died from carbon emissions and its effects than from COVID-19 during the same period.

To those of us in the West, those harmful effects have seemed far off, an affliction of poor countries. Meanwhile, we can rely on clean water flowing out of our taps and easily pour gasoline into our SUVs without any obvious effects. But now climate change is a more immediate threat as water restrictions are imposed in the drought-stricken American West, people lose their homes to wildfires, insurance companies pay out more in storm and flood damage, and farmers struggle to find irrigation water for crops. It's not just about polar bears anymore. Climate change is, in military terms, a clear and present danger.

Governments have gathered at United Nations summits in Rio, Kyoto, Copenhagen, Paris, and Glasgow to pledge emission targets, but after twenty-six meetings over thirty years, those targets have not been met. The goalposts keep moving and the climate curve is still not flattening fast enough.

But there is hope.

Rather than become depressed over all the bad news coming out of environmental stories or frustrated at the glacial pace of climate action, there is encouragement from a scientific perspective. Switching to clean energy and reaching beyond net-zero carbon emissions is entirely doable with current technology. The pace of technological change is quickening as batteries evolve to the point where vehicles can travel as far as they would on a tank of gasoline, or farther, with charging stations popping up across the land. Some countries, especially in Europe and parts of Asia, are setting examples by moving to wind, solar, geothermal, and other forms of clean energy while keeping their economies strong and improving the lifestyle of their citizens. The solutions are there. It is a matter of implementing them before the climate clock runs out.

Climate knows no boundaries. Action on climate change requires the same global effort as fighting a pandemic. We have seen how good science, political will, international co-operation, corporate investment, and public acceptance can produce positive results to fight a deadly disease.

Let's do that again with energy and save even more lives.

There really are alternatives.

While on a sailing trip down the St. Lawrence River on the way to the East Coast, I managed to run into a rock by passing on the wrong side of a channel marker. (Any sailor who says they have never run aground is either lying or hasn't gone anywhere!) After the shock and terrible sound of the hull striking bottom, I immediately went below and opened a hatch in the floor to see if any water was coming in. None was visible, so I thanked my lucky stars, admonished myself for being inattentive, and continued on.

That evening, while anchored in a small, protected bay, my sleep was interrupted by the sound of the bilge pump coming on. Normally, that is not unusual because all boats take on a little water, but this time the pump was coming on again and again, and as the night progressed, the frequency was increasing. I clearly had a problem that was getting worse.

At this point I had two choices. A short-term fix would be to replace the current bilge pump with a bigger one. As long as water is pumped out faster than it is coming in, the boat will stay afloat. And while that would allow me to continue my journey, the fate of the boat would rely entirely on that pump.

The more difficult, but better solution would be to have the boat hauled out of the water, find the hole, and repair it from the outside. Fortunately, I made the right decision. It turned out that I had struck a particularly vulnerable part of the hull, one that had been struck before, and patched, by a previous owner. The rock I hit had made the patch come loose. It took two days of fibreglass work to repair the hull. My trip was delayed slightly, but it was worth it to have peace of mind for the rest of the journey.

This situation is analogous to our climate change predicament. The rising graphs showing increased carbon emissions along with rising global temperatures are like the water level rising in the bilge of a boat. You can ignore it (for a while), you can come up with short-term fixes, or you can get to the source of the problem and fix it for good.

Glossary of Energy Terms

Energy: The capacity to do work. This could be in the form of solar energy, chemical energy, thermal energy (heat), potential energy (able to fall from a height), kinetic energy (motion), or nuclear energy (within the nucleus of an atom). (Referenced throughout text)

Gigajoule: 1 billion joules, equal to 26 litres of gasoline (p. 194)

Gigawatt (GW): 1,000 megawatts or 1 billion watts. Canada's Bruce Power Generating Station generates more than 6 gigawatts of nuclear energy, enough for a million homes. A lightning bolt can produce a gigawatt of energy, which is why Doc Brown in the movie *Back to the Future* needed the power of lightning at 1.2 gigawatts to send Marty McFly back to his future home. That DeLorean was quite an energy hog! (p. 58, p. 92, p. 112, p. 177)

Joule: A metric unit of work and energy. One joule is equal to the energy transferred to an object when a force of 1 Newton moves it through a distance of 1 metre. Lift an apple off a table up to eye level and you are expending 1 joule of energy. (p. 3)

Kilowatt (kW): 1,000 watts (p. 16, pp. 64–5, p. 76, pp. 84–5, p. 94, p. 97, p. 120, p. 150, p. 259, p. 268)

Kilowatt hour (kWh): Number of kilowatts generated or used in one hour. This is usually how your electrical bill is calculated. A 1,000-watt hair dryer running for one hour will use 1 kilowatt hour of electricity. A 2,000-watt dishwasher will use that in half an hour. An average North American home uses around 1,000 kilowatt hours a month. (p. 16, p. 20, p. 31, p. 77, p. 97)

Megawatt (MW): 1,000 kilowatts or 1 million watts. A 1-megawatt generator could power more than a thousand homes. The largest wind turbine generates 15 megawatts of electricity. An average fossil fuel generating station puts out between 600 and 1,000 megawatts. (p. 34–5, p. 46, p. 62–6, p. 84–5, p. 94–7, p. 112, p. 117–18, p. 120, p. 123–24, p. 136, p. 139, p. 159, p. 174–77, p. 182, p. 208, p. 237, p. 269)

Milliwatt: One-thousandth of a watt. A laser pointer uses about 5 milliwatts. (p. 270)

Mtoe: Millions of tonnes of oil equivalent. This metric unit is used for world consumption to compare different energy sources, such as wind, solar, biomass, etc. (p. 2)

Power: Rate at which work is done, or how much work is done in a period of time. If you lift something off the floor and time how long it takes, you can calculate how much power you use. One metric horsepower is 75 kilograms raised 1 metre in 1 second. That is also 750 watts. (Referenced throughout text)

Terawatt (TW): 1 trillion watts. The total world energy consumption is about 17 terawatts. (p. 16, p. 46, p. 80)

Watt: In electric terms, 1 ampere moving through a wire at 1 volt. A lightbulb is 100 watts. (p. 16, p. 23, p. 113)

Watt hours per kilogram (Wh/kg): Number of watts produced per kilogram of energy storage device. (p. 113)

Endnotes

INTRODUCTION

1.   Chandler, D.L. "Shining brightly: Vast amounts of solar energy radiate to the Earth constantly, but tapping that energy cost-effectively remains a challenge" (26 Oct 2011) MIT News.

2.   International Energy Agency, "Global energy review" (2020) IEA.

3.   Ritchie, H, et al. "Fossil fuels" (2020) Our World in Data.

4.   U.S. Energy Information Administration, "Units and calculators explained" (12 May 2021) EIA.

5.   Smil, V. "Building the great pyramid" (Jun 2020) *IEEE Spectrum* 18–19.

6.   International Energy Agency, "World energy outlook 2021 executive summary" (2021) IEA.

7.   Yates, Z. "The efficiency of the internal combustion engine" (2002).

8.   Jacobson, M.Z. "Outdoor plus indoor air pollution from fossil fuels, biofuels, bioenergy, and biomass burning is the second leading cause of death world-wide, and a 100 percent WWS world will eliminate most of these deaths" (2020) *100% Clean, Renewable Energy and Storage for Everything* 449.

SOLAR POWER

1.   Navarro, M.B. "Interview with author" (2021).

2.   Thoubboron, K. "Concentrated solar power: What you need to know" (4 Jul 2019) EnergySage.

3.   Woodford, C. "Solar cells" (22 Jan 2022) Explain that Stuff.

4.   Chu, E, et al. "A brief history of solar panels" (2022) *Smithsonian Magazine.*

5.   Matasci, S. "How solar panel cost and efficiency have changed over time" (8 Jan 2021) EnergySage. LaFond, et al. "Solar PV module prices" (2017) Our World in Data.

6.   Four Peaks Technologies, "Solar efficiency limits" (2011) Solar Cell Central.

7.   Oxford PV, "The Perovskite-Silicon tandem" (2018) Oxford PV.

8.   Harkness, M, et al. "Regional assessment: Stateline solar farm project. Technical report prepared for desert stateline, LLC" (27 Jul 2012) NatureServe.

9. NREL Transforming Energy, "Solar everywhere: NREL pioneers the future of photovoltaics" (13 Jan 2021) NREL.

10. UN Environment Programme, "Why do forests matter" (n.d.) UNEP.

11. Tesla, "Megapack" (2022) Tesla.

12. Power Technology, "Pavagada solar park, Karnataka" (15 May 2020) Power-technology.

  Green Energy Institute, "100 megawatts per day: Solar power on the rise" (3 Nov 2014) Lewis & Clark Law School.

13. Gupta, U. "Solar arrays on canals" (10 Mar 2021) *PV-Magazine*.

14. Energy Watch Group, et al. "Global energy system based on 100% renewable energy" (4 Dec 2019) Energy Watch Group.

BIOFUELS

1. U.S. Department of Energy, "Biofuels & greenhouse gas emissions: Myths versus facts" (28 Jul 2011) Energy.

2. Archer-Daniels-Midland Co, "Fuel" (2022) ADM.

3. European Pellet Council, "A unique biomass fuel" (2022).

4. WPAC, "The role of wood pellets in meeting climate change goals" (29 Sep 2020) *Canadian Biomass Magazine*.

5. Edwards-Evans, H, et al. "UK's drax completes acquisition of Canadian wood pellet company pinnacle" (13 Apr 2021) S&P Platts.

6. Economic Research Service, "Feedgrains sector at a glance" (28 Jun 2021) U.S. Department of Agriculture.

7. U.S. Energy Information Administration, "Monthly biofuel production report: With data for Dec 2020" (Feb 2021) U.S. Department of Energy.

8. ETIP Bioenergy, "HVO/HEFA" (2022) ETIP bioenergy.

9. U.S. Energy Information Administration, "U.S. inputs to biodiesel production, table 3" (Feb 2021) U.S. Department of Energy.

10. Bioenergy Technologies Office, Office of Energy Efficiency & Renewable Energy, "Algal logistics" U.S Department of Energy.

11. Rafa, N, et al. "Strategies to produce cost-effective third-generation biofuel from microalgae" (1 Sep 2021) Frontiers in Energy Research.

12. Le Feuvre, P. "Are aviation biofuels ready for take off?" (18 Mar 2019) International Energy Agency.

13. Le Feuvre, P. "Are aviation biofuels ready for take off?" (18 Mar 2019) International Energy Agency.

14. Cruse, R, et al. "Water use and sustainable biofuel production" (3 Apr 2019) Farm Energy Extension.

Biofuel.org.uk, "Water and biofuels" (2010) Biofuel.

15. U.S. Energy Information Administration, "How much gasoline does the United States consume?" (7 Sep 2021) EIA.

16. McConnell, Michael. "Feedgrains sector at a glance" (28 Jun 2021) United States Department of Agriculture.

17. Ruzic, D. "Biofuels and biomass: Economics of biofuels" (13 May 2019) University of Illinois: Urbana-Champaign.

18. Medrut, F. "18 John Muir quotes to deepen your connection with nature" Goalcast.

WIND POWER

1. World Wind Energy Association, "Worldwide wind capacity reaches 744 gigawatts – an unprecedented 93 gigawatts added in 2020" (24 Mar 2021) WwindEA.

2. Jaganmohan, M. "Number of active wind power turbines in Denmark 2010-2021" (3 Dec 2021) Statistica.

3. GE Renewable Energy, "An industry first: Haliade-X offshore wind turbine" (2022) GE.

4. GE Renewable Energy, "An industry first: Haliade-X offshore wind turbine" (2022) GE.

5. Siemens Gamesa Renewable Energy, "SG 12-222 DD: Offshore wind turbine" (2022) Siemens Gamesa.

Vestas, "V236-15.0 MW" (2021) Vestas.

6. TransAlta, "Wind: One of Canada's largest wind power generators" (2022) TransAlta.

7. Canadian Geographic Enterprises, "Wind energy in Canada" (2009) Natural Resources Canada.

8. Conserve Energy Future, "The comparison of horizontal and vertical axis wind turbines" (2022) Conserve Energy Future.

9. Afework, B, et al. "Betz's limit" (21 Jul 2018) Energy Education.

10. Ministry of the Environment Conservation and Parks, "Chapter 3: Required setback for wind turbines" (10 May 2019) Ontario.

11. Eveleth, R. "How many birds do wind turbines really kill?" (16 Dec 2013) *Smithsonian Magazine.*

12. Helman, C. "How green is wind power, really? A new report tallies up the carbon cost of renewables" (28 Apr 2021) *Forbes.*

OCEAN WAVE POWER

1. Institute for Integrated Energy Systems, "West coast wave initiative" (2017) University of Victoria: Faculty of Engineering.

2. Schroeder, G. "Powering the future" (2011) Voith Hydro Holding GmbH & Co.KG: Germany.

TIDAL POWER

1. Evans, S. "La Rance: Learning from the world's oldest tidal project" (28 Oct 2019) Power-Technology.

2. Unwin, J. "Potential vs. expense: Is tidal energy worth the cost?" (21 Mar 2019) Power-Technology.

3. Maine Department of Environmental Protection, "Information sheet: Regulation of tidal and wave energy projects" (2018) Maine.

ENERGY STORAGE

1. Klender, J. "Tesla Giga Berlin's 4680 battery factory is coming together" (22 Sep 2021) Teslarati.

2. Energy Vault, "Harnessing the power of gravity" (n.d.) Energy Vault.

3. ReNews.Biz, "Gravitricity explores Czech coal mine for MW-scale storage" (26 Oct 2021) *ReNews*.

4. Patowary, K. "Gyrobus: The flywheel-powered public transportation" (5 Feb 2019) Amusing Planet.

5. Energy Storage Association, "Mechanical energy storage" (2022) Energy Storage.

6. Polar Night Energy, "Technology: Solar wind and solar power as heat in sand" (2019) Polar Night Energy.

7. American Sterling Company, "How stirling engines work: An in-depth explanation" (n.d.) Stirling Engine.

8. Ambri, "Liquid metal battery" (2022) Ambri.

9. Parry, D., et al. "Australian sand: A proven, practical and inexpensive way to story energy, part 2" (13 Feb 2020) https://www.youtube.com/watch?v=uLtn8UeBUtM.

10. Hydrostor Inc, "Advanced compressed air energy storage. A unique storage solution" (n.d.) Hydrostor.

    IDTechEx, "Australia's first compressed air energy storage facility" (2020) Off Grid Energy Independence.

11. Herrington, R. "Mining our green future" (2021) 6 *Nature Reviews Materials* 456–458.

12. Guzmen, L. "Lithium sparks disputes in Chile's Atacama Desert" (16 Oct 2020) Diálogo Chino.

GEORTHERMAL ENERGY

1. Eyewitness to History, "The destruction of Pompeii, 79 AD" (1999) Eye Witness to History.

2. Cool Cosmos, "How hot is the sun?" (2013) Cool Cosmos.

3. Piesing, M. "The deepest hole we have ever Dug" (6 May 2019) BBC.

4. Pratt, S.E. "Benchmarks: March 1961: Project Mohole undertakes the first deep-ocean drilling" (6 Jul 2016) *Earth Magazine*.

5. Umino, S. "Project M2M-mohole to mantle – mantle quest thorough ultradeep ocean drilling by the deep sea scientific drilling vessel chikyu" (22 Feb 2016) 71(2) *Journal of Applied Mechanics*.

6. Canadian Geothermal Energy Association, "Where are geothermal resources located in Canada?" (2022) CanGEA.

7. IEA, "Renewables 2018" (2018) IEA.

SMALL NUCLEAR

1. World Nuclear Association, "Three mile island accident" (2020) World Nuclear.

2. Dunning, B. "Fukushima vs. Chernobyl vs. Three mile island" (14 Jan 2014) *Skeptoid*.

3. National Park Service U.S. Department of Interior, "Trinity site" (18 Sep 2017) NPS.

4. Dunning, B. "Fukushima vs. Chernobyl vs. Three mile island" (14 Jan 2014) *Skeptoid*.

5. Process Industry Forum, "The five worst nuclear disasters in history" (n.d) Process Industry Forum.

6. Process Industry Forum, "The five worst nuclear disasters in history" (n.d.) Process Industry Forum.

7. Hanania, J, et al. "Molten salt reactor" (26 Aug 2015) Energy Education.

FUSION POWER

1. ITER Organization, "Advantages of fusion" (2022) ITER.

2. ITER Organization, "What is a tokamak?" (2022) ITER.

3. Wan, X, et al. "Overview of the present progress and activities on the CFETR" (23 Jun 2017) 57 Nuclear Fusion.

4. ITER Organization, "What is ITER?" (2022) ITER.

5. Greenwald, M. "Status of the SPARC physics basis" (2020) 86 (5) *Journal of Plasma Physics*.

6. National Academics of Sciences, Engineering, and Medicine, "Bringing fusion to the U.S. grid" (2021) The National Academics Press: Washington, DC.

7. Lawrence Livermore National Laboratory, National Ignition Facility & Photon Science, "Pursing fusion ignition" (n.d.).

8. General Fusion, "The benefits of fusion" (n.d.) General Fusion.

RETHINKING OIL

9. Our World in Data, "Fossil fuel production over the long-term, United-Kingdom, 1950 to 2014" (2014) Our World in Data.

10. ProCon, "Historical timeline: History of alternative energy and fossil fuels" (14 Feb 2022) Alternative Energy ProCon.

11. ProCon, "Historical timeline: History of alternative energy and fossil fuels" (14 Feb 2022) Alternative Energy ProCon.

12. Bellis, M. "A history of the automobile: The evolution of the car dates all the way back to 1600s" (6 Jul 2019) ThoughtCo.

13. Office of Energy Efficiency & Renewable Energy, "Hydrogen production: Natural gas reforming" (n.d.) U.S. Department of Energy.

14. Office of Energy Efficiency & Renewable Energy, "Hydrogen production: Natural gas reforming" (n.d.) U.S. Department of Energy.

15. Proton Energy, "Proton's process" (n.d.) Proton Energy.

HYDROGEN POWER

1. American Rhetoric, "Herbert Morrison WLS radio (Chicago) broadcast on the Hindenburg disaster" (15 Dec 2017) American Rhetoric.

2. Wikipedia, "Hindenburg disaster" (28 Feb 2022) Wikipedia.

3. Lavo, "The Lavo green energy storage system" (n.d.) Lavo.

4. Ritchie, H. "Climate change and flying: What share of global CO_2 emissions come from aviation?" (22 Oct 2022) Our World in Data.

5. Airbus, "Airbus reveals new zero-emissions concept aircraft" (21 Sep 2020) Airbus.

CARBON CAPTURE

1. International Energy Agency, "A new era for CCUS" (2020) IEA.

2. SaskPower, "Boundary dam carbon capture project" (2022) SaskPower.

3. ZeroCO2, "Boundary dam integrated CCS project" ZeroCO2.

4. Eldardiry, H, et al. "Carbon capture and sequestration in power generation: Review of impacts and opportunities for water sustainability" (2018) 8(6) *Energy, Sustainability and Society.*

5. Shell Canada, "Quest CCS facility captures and stores five million tonnes of CO_2 ahead of fifth anniversary" (9 Jul 2020) Shell.

6. Our World in Data, "Annual CO2 emissions from cement, 2020" (2020) Our World in Data.

7. CarbonCure, "CarbonCure's carbon removal technologies" (n.d.) Carbon-Cure.

8. Lanthier, N. "Turning concrete into a sustainable building material" (14 Dec 2021) *The Globe and Mail*.

9. Solidia Tech. "The world today" (2019) Solidia Tech.

10. XPrize, "$100 million prize for carbon removal" (2020) XPrize.

11. International Energy Agency, "Carbon capture, utilisation and storage" (11 Jan 2022) IEA.

ENERGY EFFICIENCY—THE INVISIBLE POWER PLANT

1. Lucon O., et al. "Buildings." In: *Climate change 2014: Mitigation of climate change. Contribution of working group III to the fifth assessment report of the intergovernmental panel on climate change* (n.d.) Cambridge University Press.

2. Lovins, A.B. "How big is the energy efficiency resource?" (2018) 13(9) *Environmental Research Letters*.

3. Housely, "The top 10 energy efficient buildings in the world" (n.d.) Housely.

4. Lovins, A.B. "How big is the energy efficiency resource?" (2018) 13(9) *Environmental Research Letters*.

5. Lovins, A.B. "How big is the energy efficiency resource?" (2018) 13(9) *Environmental Research Letters*.

6. Traffic Injury Research Foundation, "Our road safety agenda" (2017) TIRF.

7. Freightera, "Train vs Truck – efficiency, cost, advantages & disadvantages" (2019) Freightera.

8. Ritchie, H. "Which form of transport has the smallest carbon footprint?" (13 Oct 2020) Our World in Data.

9. Mallet, R. "French train breaks speed record in Champagne" (3 Apr 2007) Reuters.

10. ACP Rail Internationa, "Bullet trains" (2020) ACP Rail.

11. Wang, O. "China plans expansion of high-speed railway equal to combined length of next 5 largest countries by network size by 2025" (20 Jan 2022) *South China Morning Post*.

12. Vantuono, William C. "Why is high-speed rail such a heavy lift?" (11 Feb 2022) *Railway Age*.

    Speck, Jeff. "Four ways to make a city more walkable" (2 Mar 2017) TED.

13. Japan RailPass, "The Japanese maglev: World's fastest bullet train" (21 Jan 2022) JRailpass.

14. Grafton LSR, "About the Bloodhound LSR Project" (Jan 2021) Bloodhound.

15. Transpod, "Calgary to Edmonton in 45 minutes: TransPod announces results of feasibility study of a hyperloop line in Alberta, confirming commercial viability and economic benefits" (25 Jun 2021) Transpod.

16. Speck, Jeff. "Four ways to make a city more walkable" (2 Mar 2017) TED.

A GREAT IDEA BUT . . .
1. Britannica, "Joseph-Michel and Jacques-Etienne Montgolfier" (6 Feb 2018) *Britannica*.

2. Solaripedia, "Solar updraft towers generate mega power" (2011) Solaripedia.

3. Wikipedia, "List of Canadian provinces and territories by area" (28 Feb 2022) Wikipedia.

4. Goddard Space Flight Center, "Webb space telescope" (n.d.) NASA.

5. NASA Langley Research Center, "Global effects of Mount Pinatubo" (15 Jun 2011) Earth Observatory NASA.

6. SkySails Group, "SkySails power technology" (2022) SkySails Group.

7. X.Company, "Makani: Harnessing wind energy with kites to create renewable electricity" (2021) X.company.

8. X.Company, "Makani: Harnessing wind energy with kites to create renewable electricity" (2021) X.company.

9. X.Company, "Makani: Harnessing wind energy with kites to create renewable electricity" (2021) X.company.

PUTTING IT ALL TOGETHER
1. WorldOMeter, "Coronavirus death toll" (1 Mar 2022) WorldOMeter.

2. EcoWatch, "Fossil fuels kill nearly 9 million annually, more than twice previous estimate, study finds" (9 Feb 2021) EcoWatch.

3. World Health Organization, "Climate change" (2022) WHO.

Acknowledgments

I would first like to thank my partner Jennifer for continued support and encouragement throughout this project.

Thanks to Lesley J. Evans Ogden for the research finding experts from around the world; Dr. María Bernechea Navarro, senior researcher (ARAID Foundation) at Instituto de Nanociencia y Materiales de Aragón (INMA), CSIC-Universidad de Zaragoza, Spain; Dr. Wim Sinke, professor of Photovoltaic Energy Conversion at TNO, Netherlands Organization for applied scientific research; Dr. Jack Sadler, University of British Columbia and International Energy Agency Bioenergy Task 39 BC Smart program; Dr. Katherine Dykes, senior researcher at ETU Wind Energy in Denmark; Wayne Oliver, operation supervisor for western Canada Trans Alta; Dr. Curran Crawford, Integrated Energy Systems, University of Victoria; Dr. Michael Ross, industrial research chair in Northern Innovation at Yukon University; Dr. Arvind Rajendran at the University of Alberta; Dr. David Layzell, professor and director, Canadian Energy Systems Analysis Research (CESAR) initiative, at the University of Calgary; Amory Lovins, Rocky Mountain Institute; Dr. Stephanie Diem, astrophysicist, University of Wisconsin Madison; Dr. Carolyn Kuranz, associate professor of Nuclear Engineering and Radiological Sciences at University of Michigan; Dr. Catherine Hickson, vice president of Geothermal Canada; Dr. Anthony Perl of Urban Studies and Political Science at Simon Fraser University; Dr. Todd Allen, Glenn F., and Gladys H. Knoll, department chair of Nuclear Engineering and Radiological Sciences at University of Michigan, Ann Arbor; Dr. Lekelia (Kiki) Jenkins, associate professor in the School for Innovation in Future Society at Arizona State University; and Dr. Bryson Robertson, University of Oregon director, Marine Energy Center.

I would also like to thank John Pearce of Westwood Artists for bringing my ideas to market, Eleanor Gasparik for meticulous copy editing, Marcia Gallego for a thoughtful proofread, and Nick Garrison for guidance to final production.